Understanding Charles Darwin

The legend of Charles Darwin has never been more alive or more potent, but by virtue of this, his legacy has become susceptible to myths and misunderstandings.

Understanding Charles Darwin examines key questions such as: What did Darwin's work change about the world? In what ways is "Darwinism" reflective of Darwin's own views? What problems were left unsolved? In our elevation of Darwin to this iconic status, have we neglected to recognize the work of other scientists? The book also examines Darwin's struggle with his religious beliefs, considering his findings, and whether he was truly an atheist.

In this engaging account, Peterson paints an intimate portrait of Darwin from his own words in private correspondence and journals. The result is the Darwin you never knew.

Erik L. Peterson is an award-winning Professor of the History of Science and Medicine at the University of Alabama (USA), and a co-host of the podcast "Speaking of Race." Erik researches the conceptual foundations of genetics, evolutionary biology, and anthropology, and is especially interested in the persistence of race science. His book, *The Life Organic: The Theoretical Biology Club and the Roots of Epigenetics* (2017), told the forgotten story of British scientists who discovered epigenetics before the Second World War – 70 years before it revolutionized American biology. He also co-authored *A Deeper Sickness* (2022), a daily history of 2020, which critics have called "harrowing" and a call for a "national reckoning."

The *Understanding Life* series is for anyone wanting an engaging and concise way into a key biological topic. Offering a multidisciplinary perspective, these accessible guides address common misconceptions and misunderstandings in a thoughtful way to help stimulate debate and encourage a more in-depth understanding. Written by leading thinkers in each field, these books are for anyone wanting an expert overview that will enable clearer thinking on each topic.

Series Editor: Kostas Kampourakis http://kampourakis.com

Published titles:

Understanding Evolution	Kostas Kampourakis	9781108746083
Understanding Coronavirus	Raul Rabadan	9781108826716
Understanding Development	Alessandro Minelli	9781108799232
Understanding Evo-Devo	Wallace Arthur	9781108819466
Understanding Genes	Kostas Kampourakis	9781108812825
Understanding DNA Ancestry	Sheldon Krimsky	9781108816038
Understanding Intelligence	Ken Richardson	9781108940368
Understanding Metaphors in the Life Sciences	Andrew S. Reynolds	9781108940498
Understanding Cancer	Robin Hesketh	9781009005999
Understanding How Science Explains the World	Kevin McCain	9781108995504
Understanding Race	Rob DeSalle and Ian Tattersall	9781009055581
Understanding Human Evolution	Ian Tattersall	9781009101998
Understanding Human Metabolism	Keith N. Frayn	9781009108522
Understanding Fertility	Gab Kovacs	9781009054164
Understanding Forensic DNA	Suzanne Bell and John M. Butler	9781009044011
Understanding Natural Selection	Michael Ruse	9781009088329
Understanding Life in the Universe	Wallace Arthur	9781009207324
Understanding Species	John S. Wilkins	9781108987196

Understanding the Christianity–Evolution Relationship	Michael Ruse	9781009277280
Understanding Living Systems	Raymond Noble and Denis Noble	9781009277365
Understanding Reproduction	Giuseppe Fusco and Alessandro Minelli	9781009225939
Understanding Charles Darwin	Erik L. Peterson	9781009338592

Forthcoming:

Understanding Obesity	Stanley Ulijaszek	9781009218214
Understanding Creationism	Glenn Branch	9781108927505
Understanding the Nature–Nurture Debate	Eric Turkheimer	9781108958165

Understanding Charles Darwin

ERIK L. PETERSON
Department of History, University of Alabama

CAMBRIDGE
UNIVERSITY PRESS

CAMBRIDGE
UNIVERSITY PRESS

Shaftesbury Road, Cambridge CB2 8EA, United Kingdom

One Liberty Plaza, 20th Floor, New York, NY 10006, USA

477 Williamstown Road, Port Melbourne, VIC 3207, Australia

314–321, 3rd Floor, Plot 3, Splendor Forum, Jasola District Centre,
New Delhi – 110025, India

103 Penang Road, #05–06/07, Visioncrest Commercial, Singapore 238467

Cambridge University Press is part of Cambridge University Press & Assessment,
a department of the University of Cambridge.

We share the University's mission to contribute to society through the pursuit of
education, learning and research at the highest international levels of excellence.

www.cambridge.org
Information on this title: www.cambridge.org/9781009338592

DOI: 10.1017/9781009338608

First published 2023

A catalogue record for this publication is available from the British Library.

*A Cataloging-in-Publication data record for this book is available from the Library of
Congress.*

ISBN 978-1-009-33859-2 Paperback

"Erik Peterson's book on Charles Darwin introduces us to Darwin the scientist, Darwin the family man, Darwin the silicone-rich member of British upper-middle-class society, at a time when the Empire was at its peak. Never before have I got to know Charles Darwin more intimately and fully; never before did I realize that such knowledge is absolutely vital to understand the revolution associated with Darwin's name. Highly recommended."

Michael Ruse, University of Guelph, Ontario, and Florida State University

"Erik Peterson has given us a fresh, deeply informed, and engaging portrait of Darwin and his revolutionary scientific achievements. Peterson also succeeds admirably in placing Darwin in historical context, both within British society and among his gentlemanly scientific peers. At the same time, this eminently readable account dispels a series of myths and misunderstandings about Darwin's thinking and influence. Of special interest is Peterson's masterful account of Alfred Russel Wallace's independent discovery of the theory of natural selection, together with the various ways that Darwin and Wallace differed in their evolutionary thinking. Altogether, a captivating and richly informative read."

Frank J. Sulloway, University of California, Berkeley

"This well-written volume unpacks a host of misunderstandings about Darwin. In attacking the pedestal that many twentieth-century biologists erected, Peterson provides a more balanced view of Darwin while also highlighting commonly overlooked contributions from others. The Captain of the *Beagle*, Robert FitzRoy, has been portrayed as a Bible-thumping creationist arguing against Darwin's revolutionary insight, but Peterson re-emphasizes how FitzRoy helped stimulate adaptive explanations. The ornithologist John Gould, if he is mentioned at all, is termed Darwin's bird identifier, but Peterson points out that it was Gould, not Darwin, who focused on the significance of the Galapagos beaks. And Peterson's own detailed research on the history of eugenics makes him the perfect foil for the myth that the Holocaust grew out of Darwin's theory. If you want to see Darwin, warts and all, this book takes you there."

Jim Bindon, The University of Alabama

"In this delightful book, Erik Peterson explodes numerous fairytales about Darwin's life and influence. Amongst these is the fiction that Darwin was a solitary genius. Peterson's corrective is a story alive with numerous people, many of them now forgotten, who played diverse roles in making Darwin the man he was. Peterson's prose sparkles; it is conversational and engaging."

Elliott Sober, University of Wisconsin–Madison

"This crisp account of Darwin, warts and all, shows him finding his distinctive voice among earlier evolutionists, including his grandfather Erasmus, standing his ground between friends who would pull him toward creationism and others who would push him toward atheism, and after his death recruited to this day for a host of dubious causes. Built on thorough knowledge of the extensive archival material and current scholarship, *Understanding Charles Darwin* will be an eye-opener for students and scholars alike."

David Depew, University of Iowa

"What do we really know about Charles Darwin, *the man*? In this engaging account, Peterson uncovers the *true* story behind one of history's most iconic and mythologized scientists. *Understanding Charles Darwin* dismantles several misunderstandings lodged into our collective consciousness, resulting from what Peterson terms 'the Darwin industry.'"

Iris Clever, University of Chicago

To my twigs, G & W: resist working for the clampdown

Contents

Foreword   *page* xiii
Acknowledgments   xv

Introduction: The Legends of Charles Darwin   1

1 The Evolutionary Darwins, 1794–1835   5
The Leisure Class   11
Captured by C. Darwin, Esq.   18
The Luckiest Guest   21

2 The Truth About Atolls   31
Descent, with Some Modification   32
The First Evolutionists   34
The Long Evolutionary Vision   41
The Man Yelling "Stop!"   43
Train in Vain   46

3 London Calling, 1836–1842   51
Career Opportunities   53
The Right Profile   58
"I think..."   62

4 Darwin–Wallaceism   67
The Contender   71
Surprise!   78

	In-Betweeners	83
	Always the Bridesmaid	87
	RIP Wallaceism	94
5	**"[T]his view of life, with its several powers"**	**97**
	On the Origin of Species by Means of Natural Selection, or the	
	Preservation of Favoured Races in the Struggle for Life (1859)	99
	The Diagram	108
	Problems	112
	The Variation of Animals and Plants Under Domestication (1868)	114
	The Descent of Man; and Selection in Relation to Sex (1871)	118
	The Expression of the Emotions in Man and Animals (1872)	122
	Darwin's Darwinism	124
6	**Saint Charles's Place**	**126**
	Darwin Never "Converted"	128
	The Darwin I Never Knew	132
	The Brush with Atheism	139
	Decisively Undecided	142
	What Child Is This?	144
7	**The Struggle Is Real**	**147**
	The *Allmacht*	148
	Survival of the . . .?	152
	"Let the one with understanding solve the meaning of the number	
	of the Beast" (Rev. 13.18; NLT)	158
	The Architects of Ruin	162
	Concluding Remarks	**170**
	The Legend Machine	170
	Summary of Common Misunderstandings	179
	References	182
	Figure Credits	198
	Index	200

Foreword

Another book on Charles Darwin? Yes, and very different than anything you have previously read about him. In this fabulous book, Erik Peterson simultaneously fulfills two tasks with success. The first is to address some common misunderstandings about Charles Darwin, someone most people have heard of without reading what he had to say. Peterson takes us on an insightful journey through Darwin's writings to reveal what he actually thought about evolution, heredity, race, religion, and a lot more. But Peterson's success depends also on the second of the tasks he successfully fulfills: the clearest and richest presentation of Darwin's own influences and background, which were crucial for the development of his theory. Erasmus Darwin, Robert Grant, Harriet Martineau, Alfred Russel Wallace, and many others come splendidly alive in Peterson's book, which helps us better understand why, how, and when Darwin developed his theory. What we get in the end is a rewarding, authentic picture of one of science's "greatest heroes." Read this book and your understanding of who Charles Darwin was and what he did will never be the same again; it will be as clear and accurate as it could be. Erik Peterson has produced a gem that ought to be read widely and celebrated as a great achievement.

Kostas Kampourakis, Series Editor

Acknowledgments

Any Darwin book enters a crowded forest. Many thanks to the stately oaks of Darwiniana who, through their time, talent, effort, and wisdom, let this sapling share a spot of sunlight: Janet Browne, Greg Radick, Marsha Richmond, Michael Ruse, Phillip R. Sloan, Frank Sulloway, and John van Wyhe. Archivists at Cambridge University Library, the British Library, and the Wellcome Institute assisted immensely with preparing the ground. Kostas Kampourakis patiently cultivated the project from mere seedlings. Brant Cook, Brian Estabrook, and L. J. Weaver stripped dead wood from early drafts. At Cambridge University Press, Olivia Boult, Jess Papworth, Aleksandra Serocka, Jenny van der Meijden, skilled copy editor Lindsay Nightingale, and the entire production team expertly pruned and cared for this tree. Broken branches of interpretation, stubborn acorns of attribution, and garbled arboreal metaphors are almost certainly my fault.

Introduction: The Legends of Charles Darwin

The stone is *still there* in the garden. That's what gets me. It's not the house itself – houses decay slowly and can be preserved pretty easily, especially in Britain where even an eighteenth-century country house is not "old." It's not even the tree behind the house, alive when Charles Darwin still lived in his Down House, now propped up by guywires against inevitable collapse as a kind of totem of the great naturalist's existence. If you leave the rear exit, the one that takes you to Darwin's preserved greenhouse and the stunning flora on a pretty path lined in that particular English way of making the perfectly manicured seem somehow "natural," you might glance to the left and see behind a small iron fence a one-foot-wide stone. A round mill stone or pottery wheel, it was, or appears to have been. And through the stone's center hole protrude two short metal bars, patinaed teal with age. Given its supposed duration in this location, it's easy to imagine the stone disappearing under the turf. That was, indeed, the intended trajectory of the original stone when it was laid there in the 1870s, not long before Darwin's death in 1882. This one is a replacement, which is carefully lifted and leveled every so often. The always green lawn is cropped short around it, since that stone is absolutely meant to be seen.

It was a kind of investigation, originally. An earthworm experiment conceived of by Francis Darwin, the recently widowed third son of Charles, fifth child overall. It's maintained there as a monument to Charles Darwin, a man who more than any other stands for the life sciences. Ironically, both father and son Darwin believed it would have been covered up by now, part of the inevitable process of bioturbation – worms (other creatures, too, but they were writing a book about earthworms, Darwin's last book) excavating soil, turning it over, bringing it to the surface, and things like large flat stones gradually being subsumed into the earth.

Here's the thing, though: in an era when compound microscope technology developed rapidly, when cell theory swept through the multinational life sciences community, when "germs" were gradually replacing explanations of humors and miasmas for health and disease, when chromosomes had been witnessed, when investigators of the patterns of heredity published scores of books – and in an era when peer-reviewed scientific journals published by scientific societies shared all of these new knowledges between a growing cadre of professional biologists who increasingly had to beg for money from wealthy benefactors and governments to continue their work – Charles Darwin did few laboratory experiments.

Sure, he fiddled. He floated seeds in salty water. He bred pigeons and skinned them, comparing skeletons. He let vines twirl in the sun and measured that. He fertilized lots of orchids and fed insectivorous plants. And given the way natural history was done in that age, all those things were good enough. But looking forward just a few decades to the days of randomized trials and arrays of test tubes, what he did looks, well, primitive. He couldn't set up controls, didn't have a microtome, never made microscopy slides, had no idea how to use chemical dyes, did not write grant applications, and could not compute statistics – all standard stuff in a late-nineteenth-century biology laboratory. What Darwin did in his back garden was observe. Plants, mostly. And let's not downplay this: he was the consummate gentleman naturalist observer.

That might be fine, except that he considered himself a *geologist* for a good part of his career – paradoxical since he's the only biologist with an internationally recognized day in his honor. Darwin Day is 12 February, his birthday. That day is used to promote the life sciences, anthropology, really anything that has to do with the study of evolution in any discipline. And that's curious. There is no equivalent day for, say, Curie in chemistry, Faraday in electricity, Herschel in astronomy, Lyell in geology, and so on. In Anglo-American culture, we usually reserve named days for political or military figures or heroes with such tremendous courage and sociocultural importance

that a "day" seems barely adequate – people like Dr. Martin Luther King, Jr.,
say, or Mahatma Gandhi.

It's not just Darwin Day, either. In the United Kingdom, his hoary visage
appeared on the £10 note for years; in the United States, he has an anti-
award named after him for accidental death via exceedingly stupid behavior.
There are Darwin figurines and bobble-heads, Darwin coloring books and
cookbooks, and, of course, there's a whole school of thought known as "Social
Darwinism," once publicly advocated by the wealthiest man on the planet,
Standard Oil's John D. Rockefeller, Jr., at least according to legend. (There's no
Rockefeller Day, either.)

Let's face it: Charles Darwin, a man who actively avoided the public eye, who
felt more comfortable staring at barnacles in his home office than at any
scientific society meeting, has become a kind of secular saint – a bearded,
wizardly face of the life sciences in general and of evolutionary biology
particularly. For that reason, we preserve his library, his study, his house,
a tree behind that house, and a rock set up in his garden to measure the dirt
moved by earthworms. People like me travel a good distance to look at his
desk, his makeshift lavatory, his snuff box, and walk the "sandwalk" circuit
that he traversed daily through the woods behind his garden. Or we board
a ship to the Galápagos Archipelago west of South America and imagine him
scrambling up the barren volcanic shores among thousands of crabs and
iguanas. Both pilgrimages are buoyed by hopes that we will catch the faintest
hint of the man and his great ideas. How that transformation occurred – from
homebody stooping over barnacles and pollenating orchids in his back garden
to secular saint worthy of statues, museum displays, and brief biographies in
every undergraduate biology textbook – is itself an interesting story that I will
briefly touch on in the final chapter.

What I want to explore in the rest of the book is a different sort of mystery. How
could someone so well known, a scientific icon, really, be so often misunderstood? The resolution to this mystery, it turns out, has a lot to do with what
Darwin *represents* rather than what he actually said. In this book (Chapters 1,
3, and parts of 4 and 6), I unpack small parts of his biography to address five
widely repeated misunderstandings, not only of the man Darwin in particular,
but of what he intended to convey in his publications and what has been made

of them since. Chapter 2 addresses the misconception that Darwin discovered evolution on the Galápagos Islands; in Chapter 4, that Alfred Russel Wallace independently arrived at the same evolutionary theory as Charles Darwin. A third misconception, that Darwin's Big Idea was merely the process of natural selection, is addressed in Chapter 5. In Chapter 6, I tackle the thorny issue of Darwin's religious beliefs and the misconception that he advocated atheism; finally, in Chapter 7, the misconception that Darwin's theories pointed directly to the death camps of the twentieth century. Tying it together, I finish with an exploration of the "Darwin Industry" that created, and attempts to tear down, the legends accreting around this one man.

With someone attached to such a voluminous body of scholarly biography and popular legendarium, I could hardly attempt to say anything completely new. My goal, instead, is to tug on the old man's beard a bit, to scratch at his ideas, to peck at his words, and see if we could get a smidgen clearer about his own message to the world as he wrote it in a large number of books and hundreds of letters from the 1830s through the end of his life in 1882 at the not-that-old age of 73. That means I'll need to go beyond the mortal Darwin just a bit to address the construction and use of Darwin-*ism* in the twentieth century.

With any luck, this book will wipe away some of the dust accumulating over the old man's image. And just maybe I can convince you that the real story of Darwin is really a story about so many other people who were *not* Darwin.

1 The Evolutionary Darwins, 1794–1835

Transmutation. "*Evolutio*," if you wanted to be fancy and Italian about it. Whatever you want to call it, the grand unrolling of one type into another, connecting all living things into a single tree of life was all the rage among the society gentlemen. James Burnett, Lord Monboddo, an influential Scottish judge in the 1700s, had said shocking things about it. Monboddo's metaphysics separated humans from brutes by only the thinnest slice of cognition. And imagine how he scandalized the chattering classes when, according to rumor anyway, he suggested perhaps tails even lingered, dangling from the spinal cords of the underdeveloped. They called him an "eccentric," a fusty, argumentative judge and a voracious reader. Perhaps too learned – genius and madness, you know.

But mostly, Monboddo used this advocacy for monogenism (the idea that all humans, no matter their external appearance or race, descend from a single ancestral source – and not necessarily a human one if you go back far enough) to poke at Henry Home, Lord Kames, his intellectual sparring partner. Yet even Kames agreed that humans had once been primitive and had changed, grown, developed, *evolved* (again, if you wanted to show off and use the Latinate term for it). Granted, Kames insisted that somewhere back in the mists of time, and less time than Monboddo insisted upon, all human races had their own, independent, non-related, quite separate origin story. He was a polygenist to Monboddo's monogenism.

This is perhaps why Darwin got involved. Not Charles Darwin. We're talking about his grandfather, Erasmus Darwin, MD (1731–1802; Figure 1.1). Polygenism undergirded slavery. Different races, different *species*, no

Figure 1.1   Erasmus Darwin (grandfather of Charles).

scientific reason standing against members of one race owning or exterminating another. And Kames was totally fine with that. Polygenism suggested that of the several races – by then they'd settled on four or five – Caucasians ruled the others. That was the natural way of things. Science had shown it.

Dr. Erasmus Darwin, though, supported abolishing the slave trade, like his old friend Josiah Wedgwood (1730–95), whose renowned pottery workshop pounded out the medallion promoting the Society for Effecting the Abolition of the Slave Trade as well as all the fancy Wedgwood plates and cups adorning the homes of the *nouveau riche*. Because of Wedgwood, the plea "Am I Not a Man and a Brother?" dangled from wrists and adorned hairpins of the fashionable across the Empire until, finally, Member of Parliament for Yorkshire and ardent Anglican, William Wilberforce, overwhelmed pro-slavery opposition to ban the Transatlantic Slave Trade in 1807. Erasmus Darwin and Josiah Wedgwood were members of the Lunar Society of Birmingham, so called because its now-famous members – also including Matthew Boulton, Benjamin Franklin, Joseph Priestley, and James Watt – preferred to return home from their meetings under the light of a full moon.

They hung together in the 1770s and 80s in part from the close relationship between Erasmus Darwin and Josiah Wedgwood, cemented in 1796 when Dr. Darwin's daughter would marry Wedgwood's son. But what if all humans were related, truly, down to their primate roots?

Erasmus Darwin teased evolutionary ideas in a medical work, *Zoonomia; or the Laws of Organic Life* (1794–1796), a text that contained in embryonic form much of what would be considered evolutionary theory for the next 50 years. Dr. Darwin started with the realization that the Earth was very old – a concept he learned from his father, Robert Darwin of Elston, who procured a fossil, later identified as a plesiosaur, and donated it to the Royal Society of London. Fossils prove that living things have had millions of years to adapt, said Erasmus Darwin. Any competent naturalist can see the "great similarity of structure which obtains in all the warm-blooded animals, as well quadrupeds, birds, and amphibious animals, as in mankind." Examine the paws of a mouse, the wings of a bat, the feet of an elephant, or the flipper of a fossilized plesiosaur and "one is led to conclude that they have alike been produced from a similar living filament." Examine embryonic development and a careful observer could witness with their own eyes the markers of deep history. Common descent with modification – many, if not all, organisms related to one another over the vast expanse of time.

In some cases, Dr. Darwin pointed out, organisms, as they descended from earlier less specialized stock, "acquired hands and fingers, with a fine sense of touch, as in mankind." But other organisms evolved out of the same starter package "claws or talons, as in tygers [*sic*] and eagles" or "toes with an intervening web, or membrane, as in seals and geese" or "cloven hoofs [*sic*] as in cows and swine; and whole hoofs [*sic*] in others, as in the horse." Even in birds, Dr. Darwin insisted, one could see the evidence of common descent, in this case shared with mammals. Over millions of years, similar hard parts (e.g., bones, teeth, horns, beaks) in an overall body-plan were tweaked just so to produce very different organs with different functions: "in the bird kind this original living filament has put forth wings instead of arms or legs, and feathers instead of hair. In some it has protruded horns on the forehead instead of teeth in the fore part of the upper jaw; in others, tushes [short tusks] instead of horns; and in others, beaks instead of either." He saw the same evidence of evolutionary descent in plants, as he publicized in 1791 in *The Botanic Garden*.

Truly, as Dr. Darwin proclaimed in *Zoonomia*, "the whole is one family of one parent." Evolution, transmutation, unrolling, common descent, whatever – it was all part of God's master plan, a law from the start, like gravity, according to Erasmus Darwin.

In the posthumously published poem *The Temple of Nature; or, The Origin of Society*, published in 1803, Erasmus Darwin went even further in detailing his evolutionary vision. And because it was presented poetically, modern scholars like to dismiss it. But take a look; Dr. Darwin packed the margins around the poem full of detailed footnotes and then tacked on multiple appendices, totaling almost 200 pages. This was no flight of fancy. Dr. Darwin laid out a detailed evolutionary vision several years before his better-known grandson was even born. He even beat the other "father of evolution," Jean-Baptiste Lamarck (more on him later).

Humans, Dr. Darwin said in *Temple of Nature*, descended from ancestors who probably originated in the place we now call Syria. This conjecture belonged to eccentric Monboddo, who had deduced it from studying human languages. But those human ancestors, Dr. Darwin insisted, had their own primate ancestors: "one family of monkeys on the banks of the Mediterranean; who accidentally had learned to use the *adductor pollicis*, or that strong muscle which constitutes the ball of the thumb, and draws the point of it to meet the points of the fingers...." With their advanced thumbs, accidentally acquired and passed on through the generations, these Mediterranean monkeys began to pick up all the other specialized things we humans do. Eventually, they crossed some sort of a line; they became us. This, explained Dr. Darwin, was the descent of man.

Yet he speculated that this evolution of humanity by passing on acquired characteristics was just a more specialized case of the process beginning at the beginning, when heat and water gave birth to the first cells: "Nursed by warm sun-beams in primeval caves, Organic Life began beneath the waves." And here's what I find especially interesting. Very rough outlines of most aspects of evolutionary theory that we study today were already present in Erasmus Darwin's work. For instance, he realized that early living creatures must have had the power to lock-in some of their common experiences through heredity, but that they also varied substantially generation to generation. Variation likely came as a result of different environmental conditions including the climate, food, their use-and-disuse of parts to get at food and to escape death, and even disease: "The clime unkind, or noxious food, instills / To embryon nerves hereditary ills; / The feeble births acquired diseases chase, / 'Till Death extinguish the degenerate race." He even had a theory for the genesis of sexual reproduction itself. Sex originated in order to combat disease, Dr. Darwin speculated. "As the sexual progeny of vegetables are thus less liable to hereditary diseases than the solitary progenies," Erasmus Darwin reasoned in *The Temple of Nature*, "so it is reasonable to conclude, that the sexual progenies of animals may be less liable to hereditary diseases, if the marriages are into different families, than if into the same family. . . ." This, interestingly, is a leading theory of the origins of sex today.

Moreover, once differentiated into male and female sexes, Erasmus Darwin saw yet another mechanism for biological diversity: competition for mates. Antlers, tusks, showy plumage and the like all served to attract females and ward off competing males – what we now call "sexual selection." We call it that because Erasmus's grandson called it that almost three-quarters of a century later in the book *The Descent of Man; and Selection in Relation to Sex*, published in 1871 .

Erasmus Darwin even anticipated what Charles Darwin would write in the book that followed a year after the *Descent of Man*, entitled *The Expression of the Emotions in Man and Animals*. To those who objected that far more complex behaviors – language, for instance, or tool use – could not be a result of evolution, Dr. Darwin responded by highlighting how animals imitate sounds and behaviors of each other just like humans. Of course, our sound and behavior imitations are, to us anyway, more faithful, more nuanced, more sophisticated. But that basic process of learning by observation, trial, and error, much like evolution itself, Dr. Darwin suggested, can be best described as imitation with small deviations – a kind of common descent with modification visible in psychology as well as biology.

I think it's fair to say many ideas that would reappear in the works of Erasmus's better-known grandson Charles already hovered in Darwin family literature and thought. It's difficult to ascertain, however, how directly these traits passed down to grandson Charles Robert Darwin. Given the relationship between Erasmus and his fourth child, Robert Waring Darwin (1766–1848), perhaps not all that much. Accounting is inexact, but Erasmus Darwin fathered as many as fifteen children with as many as four women. After his first wife, Mary "Polly" Howard (1740–70) died when fourth-child Robert was only four years old, Erasmus hired a governess, seventeen-year-old Mary Parker (1753–?), to care for Robert. From 1771 to 1774, Erasmus fathered three more daughters, two with Mary Parker (1772 and 1774), and a third supposedly with a woman named Lucy Swift in 1771, before marrying again, this time to the newly widowed and quite wealthy Elizabeth Pole (1747–1832). Erasmus and Elizabeth bore seven additional offspring, including Frances Ann Violetta Darwin (1783–1874), the mother of Francis Galton (1822–1911); after attacking his cousin Charles Darwin's concept of inheritance in the 1860s and 1870s, a concept borrowed in part from Erasmus Darwin's works, Galton went on to coin the word "eugenics" in 1883. Convinced by Galton and his followers, German and American eugenicists would sterilize hundreds of thousands of men and women in the name of eugenics in the years leading up to the Holocaust, though *that* event had nothing to do with a Darwin, as we will see later.

Josiah Wedgwood had promised Erasmus that his daughter, Susannah "Sukey" Wedgwood (1767–1817), would marry Robert Waring Darwin as soon as Robert had made something of himself. By 1787, 21-year-old Robert had completed medical training both at the prestigious University of Edinburgh and then at Leiden University in the Netherlands. A year later, he was elected Fellow of the Royal Society on the basis of a medical dissertation some historians suspect Erasmus helped research. Robert, now a physician, returned to the ancestral home in Shropshire to take patients. Rather than biology or medicine like his father, however, Robert loved financial investing. He dumped capital into canals and a highway, and they paid off. So, by 1795, Josiah Wedgwood gave the nod to a well-funded Robert Darwin and Sukey Wedgwood engagement. When the old potter unexpectedly gave up the ghost soon after, £25,000 (roughly £2.3 million today) went to Sukey. In other words, the newlywed Darwin-Wedgwoods began married life quite comfortably.

Grandfather Erasmus died in 1802. Robert Waring continued to practice medicine for the well-to-do. The Darwins built a large, three-storey brick

Figure 1.2 "The Mount," Shrewsbury (c. 1860), Charles Darwin's boyhood home.

Georgian house on the southern bank of the River Severn in Shrewsbury, not far from the Welsh border, and called it "The Mount." It made a respectable, soft nest for their brood (Figure 1.2).

The Leisure Class

Charles Robert Darwin (1809–82) was the name Susannah and Robert gave to their second son when he was born there on 12 February. By coincidence, that same day Nancy Hanks Lincoln gave birth to her second child, Abraham (1809–65), thousands of miles west in much less opulent surroundings in rural Kentucky. Both Abraham Lincoln and Charles Darwin lost their mothers at young ages – Darwin at eight, Lincoln, nine – and were left in the care of elder sisters. Both boys feared and admired their fathers; both Robert Waring Darwin and Thomas Lincoln scolded their sons for being lazy, for preferring books over hard work. Yet, after undistinguished educational trajectories, both boys would develop solid habits as they grew into men with minds sharper than most of their peers. Then – again at nearly the same moment in time, 1859–60 – Lincoln and Darwin each changed the world.

That would have been hard to guess had you met Charles Darwin in the 1820s. At the Shrewsbury School, a public school of some academic stature run by Archdeacon Samuel Butler (grandfather of the author and later frenemy of Darwin by the same name), Charles found it impossible to keep up with his Classics lessons. "Gas," his classmates called him, because he cared enough about his brother's chemistry studies to be teasable – and not much else. Charles was helping older brother Erasmus, nicknamed "Ras" to distinguish him from the distinguished grandfather, assemble chemical experiments according to those brother Erasmus was learning at Christ's College, Cambridge. In 1825, after a summer assisting his physician father, the mostly miserable Charles followed Ras to the University of Edinburgh to study medicine.

They boarded at 11 Lothian Street, just a couple of city blocks from the Medical School on Teviot Place and right in the heart of the great university. Charles didn't like it there either. The lanky teenager seemed the opposite of his portly, nose-to-the-grindstone father. Charles mostly skipped lectures; he found medicine stressful, pain and suffering of others acutely distressing. He took a geology class instead. But it wasn't much better; though he showed some interest in the subject, he found Regius Professor of Natural History Robert Jameson's lectures boring.

Like so many undergraduate men propped up by family wealth, he much preferred the sporting life over academic work, which in those days meant horse riding and hunting. Ras didn't fare much better at rigorous Edinburgh than Charles did and departed in the summer of 1826 to head back to Christ's College, Cambridge University. Later, Ras used those connections to construct a life of socializing into which his younger brother would dip in and out. Charles outlasted Ras at Edinburgh, but instead of consulting the scholars of renown that surrounded him, Darwin shot things. Birds, mostly. And, because he wanted to display his handiwork, he took them to a taxidermist whose shop stood at 37 Lothian Street, directly on the path from the Darwin brothers' rooms to the Medical School.

John Edmonstone was the name of the taxidermist. Or at least that was the name he adopted in Scotland. What his real name was, we may never know. He came recommended by Andrew Duncan, the still active 80-year-old former president of the Royal College of Physicians of Edinburgh. At the time, Darwin called him merely a "blackamoor" and sought him out chiefly because his services were inexpensive. Perhaps also because Darwin was just a little sheepish about remaining "most shockingly idle," at this world-class medical program, spending time reading novels and hoping to "get tipsey" by huffing "[Ni]tric oxide" with school friends, according to one of his private letters (sourced from the Darwin Correspondence Project, DCP 22). But whatever the initial reason for their interaction, Edmonstone's almost unacknowledged influence on Darwin would prove profound.

John Edmonstone had, until recently, been enslaved at a plantation south of Georgetown in what was then British (and before that, Dutch) Guiana, South America. He traveled to Scotland, perhaps around 1817, in the company of one of the largest slave owners in that part of British Guiana, Charles Edmonstone, and his wife, Helen Reid (Reid was herself supposedly a daughter of Arawak "Princess Minda"). Charles Edmonstone was also a well-known slave catcher and had led so many expeditions into the rainforest to capture and kill escaped slaves (maroons) that the colonial governors had exempted him from taxation. Whether Charles Edmonstone manumitted John or John just slipped away upon reaching the British Isles is unclear – by 1778, slavery was illegal inside Scotland, despite the fact that many prominent Glasgow and Edinburgh families gleaned their ostentatious wealth from the blood and sweat of enslaved people. John Edmonstone set up his taxidermy shop on Lothian St. in 1824, perhaps after serving as Dr. Andrew Duncan's servant, to attract university students and faculty. By all accounts, it was a shrewd choice. His method was novel; his taxidermic specimens more lifelike than anyone else's. He had learned it directly from famed explorer Charles Waterton (1782–1865), another wealthy son from plantation money.

The Watertons owned the Walton Hall island estate in Wakefield, near Yorkshire. Decades later, Charles Darwin would visit them there. The Watertons also owned Walton Hall sugar plantation and two others in British Guiana. Combined, those plantations kept enslaved over 500 African- and Native-descended men, women, and children. Though he claimed to deplore slavery and ardently support monogenism just like a Darwin or Wedgwood, Charles Waterton worked as a manager of all his family's plantations until 1812 and remained defensive about how friendly his own family was to those they held in slavery. After the men in his family who directly owned the plantations died, Waterton gradually sold them off while continuing to travel throughout Guiana and northeastern South America, returning on four occasions to collect animal specimens and visit plantations. His *Wanderings in South America, the North-west of the United States, and the Antilles in the Years 1812, 1816, 1820, and 1824*, published in 1825 and reprinted frequently over the whole nineteenth century, became a touchstone for aspirational explorers, including Charles Darwin and Alfred Russel Wallace. Waterton peppered his narrative with exciting and at times humorous accounts of encounters with dangerous jungle animals. Most famously, four natives and two enslaved men helped Waterton capture and "ride" a 10-foot-long black caiman before dissecting and stuffing it in 1820 (Figure 1.3). Charles Edmonstone owned one of the unnamed enslaved men sent on Waterton's caiman-riding adventure; Waterton was training him to stuff birds also.

Figure 1.3   Tales of Charles Waterton in Guiana in the 1820s inspired colonial collector/explorers like Darwin and Wallace.

Waterton's process of preservation may not have been revolutionary, but it preserved the color of bird feathers uncommonly well in the humid jungle and made the resulting animals displayed incredibly lifelike. Waterton-esque specimens still dot the great natural history museums of England. Mostly, his method involved cotton. Lots and lots of cotton. Wires, he said, were right out. You needed to stuff the body full, give it the look of a supple living thing, leave no voids inside as the outside dried and the skin collapsed in on itself after death. The animal needed to be quickly and very carefully dissected, skinned, cleaned of almost all internal blood and soft tissue, leaving only, at least in the case of birds, small segments of bone with which to give the preserved animal firm structure and something hard to which the taxidermist could attach thread. He also dipped the outside of the animal and cleaned a good number of its internal parts with mercury chloride ($HgCl_2$), a toxic white powder nevertheless employed in medicines and as an antiseptic until the advent of antibiotics in the twentieth century. Waterton called it "corrosive sublimate," dissolved in alcohol. This was the key. It was this mercury chloride poison solution absorbed by the specimen that killed the insects (and likely the bacteria) that would otherwise consume the animal skin and feathers after preservation. Mercury chloride ensured the specimens made it all the way back to London.

Still, this application was only the preserver's finishing touch. The taxidermist, Waterton insisted, must possess hands nimbler than a surgeon's and a "complete knowledge of ornithological anatomy"; they must also be able to sit very still with the specimen balanced upon their knee, cutting without tearing, since that was the only reliable way to smoothly raise and lower the animal toward the taxidermist while keeping feathers perfectly intact and avoiding "lassitude." By all accounts, John Edmonstone, though certainly prevented by virtue of his own skin from becoming a surgeon himself, knew how to skin a bird like a surgeon, stuff it, preserve it for museums, make it appear as if the bird would soon take flight. These skills John passed on to the wealthy, "idle" Charles Darwin, who paid one guinea for two months of bird preservation lessons in 1826. This Guianan taxidermist gifted Darwin a free primer on tropical birds and animals along the way.

Here's the pay-off to this side story. Without John Edmonstone's hands, his intimate knowledge of ornithology, and Waterton's toxic mercury chloride, fewer of Darwin's stuffed animals would have survived the slow, multiclimate ocean voyage from South America and beyond to London ornithologists, birds with beaks and features so intact that you can still see their feather coloration and beak differences today, two centuries later – traits that reveal slight modifications island to island.

Perhaps even without knowing it, young Charles Darwin also imbibed two major scientific debates swirling around the faculty of Edinburgh that directly impacted his future trajectory. The first debate raged over the age and composition of the Earth itself. Robert Jameson, who had been a vocal proponent of German "Neptunism" – the geological theory that Earth was originally covered with water, and rocks precipitated out of the water (a moderately old-Earth account that nonetheless comported with the Bible) – began converting to Scottish "Vulcanism" – the competing theory that Earth's crust forever recycles as molten rock. Vulcanism (also called Plutonism) meant no definitive "In the Beginning..." could be found in rock, given the even more immense age of the Earth conjectured. It also meant that whatever the ground appeared like now, it would one day be remade in fire. Permanence, even when Earth was concerned, flew out the window.

A second, even more dramatic controversy erupted in 1826 over an anonymous article published in the *Edinburgh New Philosophical Journal*, edited by Robert Jameson (more on this article later). The article seemingly endorsed ideas of species transformism or transmutationism coming out of France under the names of Lamarck and Geoffroy Saint-Hilaire. Painstaking historical research has recently revealed that the anonymous article was not written by Jameson. Nor was it written by the more radical "Edinburgh Lamarckian" Robert Edmond Grant, MD (1793–1874), a young comparative anatomist and burgeoning marine biologist – a sponge hunter, in other words (Figure 1.4). The anonymous article was a translation of a German article. Yet, for many years, historians regarded Grant as the most likely author because Grant read Erasmus Darwin and swallowed common descent with modification. That's important, because, when he had the chance in 1826, Grant nabbed Erasmus Darwin's grandson.

Grant met Charles Darwin along the Firth of Forth near Leith, where Grant resided. From there Grant hunted for marine invertebrates and, as well as

Figure 1.4   Robert E. Grant, one of young Charles Darwin's influences at the University of Edinburgh.

inviting him into the exciting world of sponges and sea slugs, he introduced young Darwin to the cadre of Edinburgh students interested in natural history. Darwin skipped Jameson's lectures and any others he didn't care for, but he attended Plinian Society meetings and Wernerian Society meetings in the evenings, both natural history groups full of excited men ready to pile into basement meeting rooms to argue over geology and transmutation theories. During the day, he walked with Grant or learned about the classification of plants in the Edinburgh University Museum. The Plinians inducted Darwin as a member on 28 November 1826 based on the testimony of William Alexander Francis Browne, president of the society, and he felt immediately a part of something bigger than himself. Interesting ideas floated around these meetings. Radical ideas. A fellow inductee, William Rathbone Greg, announced at the time that he would give a talk showing that "lower animals possess every faculty and propensity of the human mind" – a topic Darwin would eventually address almost five decades later. Darwin cut his teeth doing public science presentations on findings of the larval sea mat (*Flustra*) he came across on one of those expeditions with Grant. Perhaps most importantly, on more than one occasion Grant "burst forth" in praise of the transmutationist theories. It turns out Grant had absorbed the writings of Darwin's grandfather, Erasmus, around the same time he traveled to post-Napoleonic France to meet the French transmutationists firsthand. Seemingly embarrassed, Darwin admitted that he'd read *Zoonomia* (though he didn't mention the more explicitly transmutationist *The Temple of Nature*) but didn't know what to make of it.

Darwin asserted these readings and interactions meant little to him, aside from increasing his thirst to make some mark in the world of science. In fact, he claimed that the entire period in Edinburgh was basically a waste and denied that evolutionary theories were "in the air" before his major publications in the middle of the nineteenth century. But when we recall Darwin's days spent learning about birds from Edmonstone and his nights discussing transmutation with Edinburgh's radical scientists, it is hard to take his word for it.

Captured by C. Darwin, Esq.

With prodding from his daughters, Robert Darwin agreed to pull Charles out of Edinburgh, just as he had the Shrewsbury School, sending him instead to Christ's College, Cambridge, where Ras had earned his M.B. degree, a college with a reputation for being less academically rigorous than others, also attended by other Wedgwood-Darwins. There Charles would have a chance to develop into a churchman with credentials, a country parson with plenty of time for hunting sponges or shooting birds or whatever. Older Charles, looking back on it, was just as amused as you might be in learning that the intended path could have produced Rev. Charles Darwin, Church of England.

But the 19-year-old son of Darwin-Wedgwood money from "The Mount," Shrewsbury, Shropshire, was accustomed to gentlemanly, not necessarily religious, tastes, most of these acquired with minimal effort on his part. He did not study theology. He merely needed to pass two examinations to receive his B.A. When he matriculated in 1828, he brought a favorite horse from home, doubled down on partying, and attended classes even less than he did at Edinburgh in 1826–27. One of his favorite games was to shoot his rifle at a moving candle held by his inebriated friends inside his college rooms. Needless to say, the gun had no ammunition in it – the explosive cap produced enough wind to puff out the candle if he was on target. This very realistic first-person-shooter game occurred so often it led one Fellow of the College passing by to inquire why he heard Darwin constantly cracking his horse-riding whip indoors.

Beetles replaced sea sponges as his passion at Cambridge. William Darwin Fox, a slightly older second cousin through grandfather Erasmus Darwin's sister, introduced Charles to the gentlemanly pursuit of bug collecting. It turned out to be potentially the most important thing that happened to collegiate Charles Darwin. Through 1829–31, he spent an inordinate amount of time hunting beetles; not *studying* them, per se, just hunting: "It was the mere passion for collecting, for I did not dissect them and rarely compared their external characters with published descriptions...." Still, when he saw "captured by C. Darwin, esq." attached to an entry in James Francis Stephens' *Illustrations of British Entomology* (1829–32), he was ecstatic. That electric feeling of being noticed by the scientific community pushed him through much of the rest of his life.

In terms of actual education, though, it seemed not to help much. His father had already upbraided him for his laziness, and Charles never forgot the sting: "To my deep mortification, my father once said to me, 'You care for nothing but shooting, dogs, and rat-catching and you will be a disgrace to yourself and all your family.'" So, he did do some studying, eventually, if only to please his father. And eventually he did graduate as an ordinary degree student since he didn't take the more advanced *Tripos* honors exam offered by the university. In his quiet moments, Darwin regretted he "threw away" his time at Cambridge just as he had at Edinburgh.

But that's not quite fair. He may not have appreciated what he was imbibing (other than beer and cigar smoke – he lived above a tobacconist his first term at Cambridge; today it's a Boots Pharmacy). He did read, though he deviated from some of the assigned curriculum to read other books more interesting to him. William Paley, the liberal, anti-slavery, pro-American moral philosopher who once taught at Christ's College Cambridge left a particular impression on Darwin. Two Paley books, *A View of the Evidences of Christianity*, published in 1794, and *Principles of Moral and Political Philosophy*, published in 1785, served as pillars of Cambridge theology education for the entire nineteenth century. Darwin also read on his own *Natural Theology, or Evidences of the Existence and Attributes of the Deity*, published in 1802, in which Paley popularized the lasting "watchmaker" argument. Darwin later claimed he practically memorized that book.

The watchmaker argument goes something like this. Imagine you are on a long walk in an unpopulated place, and you accidentally kick a watch on the path; you immediately intuit that it is a *designed* thing (i.e., it requires a designer) because it's very complex. Funny thing is, said Paley, living organisms are way more complex than watches. So shouldn't we regard organisms as designed, too? And if making an Apple Watch takes all of a megacorporation to pull off, and given that even a simple coronavirus is much more complex than an Apple Watch – it can reproduce itself by hijacking another living thing to replicate millions of copies of itself without outside help, for instance – shouldn't we regard the designer of viruses, birds, trees, the human hand, and so on, as impossibly more advanced than any human designer?

Biologists and philosophers mock Paley's metaphorical argument today. But that's because they aren't Charles Darwin. Today's biologists are woefully unacquainted with Paley's actual motivation for writing *Natural Theology*, not to mention Darwin's reasons for reading it, retaining it, dusting it off, and sparring with it to the end of his days. It's one of the great misunderstandings about Darwin's work that Darwin intentionally killed off Paley. It's truer to say that Darwin *refined* Paley, fitting Paley's argument to new evidence. Darwin left Cambridge a pretty committed Paleyite, and not just because of the books themselves.

Two Cambridge professors, the Reverend John Stevens Henslow, a botanist, and the Reverend Adam Sedgwick, a geologist, regularly reinforced that message of natural theology to Darwin. Partly, Henslow proved convincing to Darwin simply because the professor was genuinely good, patient, selfless, and kind to the underachieving student. "My intimacy with such a man,"

Darwin reflected years later, "ought to have been and I hope was an inestimable benefit." Darwin spent so much time imbibing this message from Henslow in 1830–31 that even other Cambridge dons referred to him not as "the inveterate horse-whip cracker" nor "the beetle maniac" but as "the man who walks with Henslow." Sedgwick, too, took a special interest and invited the newly graduated Darwin to Wales in 1831 on a geologic expedition. He stitched the notion of deep time together with natural theology and gave Darwin the critical experience that made him at least somewhat more ready for the greater research task of the HMS *Beagle* voyage that would follow.

In the natural world, these natural theologians insisted, everything fit. In the natural world, everything was adapted to its environment. In the natural world, everything was, as the Psalmist said, beautifully and wonderfully made. Pain and suffering, awkward creatures, non-adaptive traits – these were only puzzles, apt to mislead, born of our ignorance of God's larger adaptative design. Charles Darwin would deviate from much of traditional religion, but he retained that conviction of the natural theologians that, despite appearances to the contrary, despite the ignorance, cruelty, and greed of human pursuits regarding rocks and plants and animals, nature worked together for good somehow.

The Luckiest Guest

In retrospect, Darwin was profoundly lucky. The *Beagle* voyage invitation, which would make him science-famous enough to have his *On the Origin of Species* taken seriously decades later, arrived at The Mount in August 1831. Henslow had just convinced Darwin to "begin the study of geology" that summer by examining rocks around Shrewsbury and coloring in a survey map. Initially the navy directed its request to Cambridge mathematician George Peacock, who in turn asked Henslow to intercede with a brother-in-law, accomplished naturalist and vicar Leonard Jenyns. But young gentleman-collector Darwin seemed less tied down than Jenyns, so Henslow passed the opportunity to the unfocused Shropshire-doctor's son. Though with limited geological experience, Darwin would play the role of ship's naturalist and companion – "a *gentleman*," Henslow stressed – to mercurial Captain Robert "Hot Coffee" FitzRoy (1805–65) on a two-year-long second expedition of the relatively small, six-gun, barque-rigged HMS *Beagle* down the eastern coast-line of South America (DCP 105).

That now well-known trip had an interesting, if less known, prelude. And, because of what occurred on that first voyage of the *Beagle*, FitzRoy needed to lean on important family connections within the Royal Navy, including expending a considerable amount of his own personal money along the way, for this second *Beagle* expedition to occur. Even given his noble lineage and pressure from his father, General Lord Charles FitzRoy (a well-known abolitionist and aide-de-camp to King George III), the navy proved quite reluctant to send HMS *Beagle* out to South America a second time. Captain FitzRoy persisted, renting an ocean-going ship himself and taking leave from the navy, so anxious was he to return. As well as sampling flora and fauna from land and sea along the southern edge of South America and completing the mapping of the difficult sea passages around the southern tip of the continent aborted during the first mission, FitzRoy aimed to return something important. Or, rather, some*one*.

In January 1830, during the first voyage of the *Beagle*, the newly minted Captain FitzRoy took onboard four Aboriginal Fuegians, initially as bargaining chips to regain a stolen whaling boat. When those negotiations failed, pur-portedly for lack of interest in a trade by the Fuegians, FitzRoy transported "Fuegia Basket" (a pre-teen girl whose Fuegian name was Yokcushlu), "James 'Jemmy' Button" (a teenage boy; Fuegian name Orundellico), "York Minster" (an adult man, perhaps 26; Fuegian name Elleparu), and "Boat Memory" (another young man, perhaps 20, and FitzRoy's favorite) to England (Figure 1.5). The English names, of course, were made up by the sailors. FitzRoy insisted they receive an initial inoculation against smallpox even before leaving South America – he knew firsthand what European diseases did to native peoples around the world. When in England, he had them immediately inoculated a second time, then set them up in a country farm-house where they could acclimate and avoid public attention and disease. Nevertheless, Boat Memory died of smallpox at the Royal Naval Hospital in Plymouth; his infection may have been the one introduced by the second inoculation. Over the next year, the remaining three received an education at the new St. Mary's Infant School in Walthamstow, a northeastern town in Greater London, chosen because the kindly instructor Rev. William Wilson

Figure 1.5   Captain FitzRoy's depiction of the three surviving Fuegians who traveled on the *Beagle*'s first and second journeys.

practiced radical "instruction by amusement" with plenty of games and exercise. Fuegia/Yokcushlu and Jemmy/Orundellico quickly took to the lessons. And, in the spring, FitzRoy presented them to King William IV and Queen Adelaide at St. James' Palace. The school relegated older York/Elleparu, on the other hand, to menial tasks around the campus. Instead, he took to young Fuegia/Yokcushlu. On their return trip to South America, they were considered "engaged" and, once back in Tierra del Fuego, married. Many years later, Darwin learned of their fates. Jemmy/Orundellico had a son, then grandchildren. Fuegia/Yokcushlu, too, made life work in her Patagonian world. Her husband York/Elleparu, however, died in an argument with another Fuegian.

Eventually, the navy relented and gave FitzRoy command of HMS *Beagle* for a return journey. It was not just a mercy-mission to return Fuegians. The Admiralty directed FitzRoy to correct Spanish charts of the whole coastline of South America. They outfitted the *Beagle* with azimuths, chronometers, and telescopes as well as the usual nautical instruments. FitzRoy should measure magnetic variation in the Earth, capture tides, currents, barometric pressures, wind speeds more precisely than "the ambiguous terms 'fresh,' 'moderate,' etc., in using which no two people agree," slopes of the land, coral reefs, instances of vulcanism, and the salt-to-fresh water transition in estuaries. This was to be an overtly scientific mission, carefully observing "eclipses, occultations, lunar distances, and moon-culminating stars" from different latitudes. He must keep records of eclipses of Jupiter's moons, especially Ganymede and Callisto. He had to capture observations of comets as well as other "remarkable phenomena" that might prove "highly interesting to astronomers."

With this long list, FitzRoy immediately requested he be permitted a geologically trained *savant*, the position Charles Darwin filled. He justified this request because of his frustratingly inaccurate compass readings during the first *Beagle* voyage, which hinted that nearby mountains contained a high degree of magnetic ore. But historians have long speculated that FitzRoy harbored other darker reasons for requesting an educated companion, and the Admiralty granting that request.

The original captain of the *Beagle*, Pringle Stokes, attempted suicide due to loneliness and the stress of command in dangerous seas at the southern tip of South America. The bullet didn't immediately kill him but stayed lodged in his head; infection painfully claimed Captain Stokes two weeks later. Already an accomplished flag lieutenant on the surveying ship HMS *Ganges*, the 23-year-old

FitzRoy found himself the new captain of HMS *Beagle*. From December 1828 to spring 1830, he got a taste of what Stokes experienced in those tempestuous southern oceans. Perhaps that alone frightened him.

But he had even more reason to worry the same fate might lie in store for him: though under suicide-watch, FitzRoy's distinguished uncle, Foreign Secretary Lord Castlereagh, managed to successfully slit his own throat with a penknife in 1822. What if depression and suicidal tendencies ran in the family? If he wondered this, FitzRoy was prescient. Depression, supposedly instigated by a reprimand from the Admiralty, pushed him to temporarily relinquish command of the *Beagle* while separated from Darwin off the coast of Chile in 1834. And, sadly, on 30 April 1865, the 60-year-old FitzRoy, who was by then a pioneering meteorologist, would finish his morning correspondence, quietly lock the door to his room behind him, and imitate his uncle Lord Castlereagh with a razor of his own.

Though only four years apart in age, the captain and his companion were not social equals. FitzRoy (Figure 1.6) possessed social status and adhered to strict

Figure 1.6   Captain Robert FitzRoy.

military discipline. Darwin appeared soft to FitzRoy; his face, especially the shape of his nose, gave it away. (Curiously, Darwin later claimed in his autobiography that FitzRoy followed Johann Kaspar Lavater (1741–1801), a Swiss Calvinist promoting materialism, much like Charles' grandfather, Erasmus.) Upon closer inspection, Darwin didn't seem exactly the experienced geologist that FitzRoy hoped for. This was one of Robert W. Darwin's objections to Charles going on the voyage in the first place – Charles didn't have the qualifications. The Admiralty never conferred an official title or salary on Darwin at all. These facts and others have led historians for a half-century to debate whether Darwin was the "official naturalist" on the *Beagle*. Perhaps Surgeon Robert McCormick was the true ship's naturalist.

Though more recent historical sleuthing suggests that indeed Darwin was ship's naturalist, surgeons often also played the role of naturalists in the British Navy. McCormick seemed especially well-suited for that task. He had studied at Edinburgh, like Darwin. Indeed, he had pursued a full year of natural history courses with geologist Robert Jameson, the professor whose classes Darwin couldn't quite work up the wherewithal to attend. (After leaving the *Beagle* voyage in 1832, McCormick would serve as ship's surgeon on the HMS *Terror*, then the HMS *Erebus* on its 1839 Antarctic expedition. Both ships, famously, would later be lost attempting to find the Northwest Passage. McCormick's assistant surgeon on *Erebus* was Joseph Hooker, a future confidant of Charles Darwin's.) But McCormick quickly came to resent being cooped up on the *Beagle*, apparently disallowed the freedom to explore he expected. He asked to be transferred from the *Beagle* less than half a year into the voyage.

Several other *Beagle* passengers assisted Darwin in collecting after McCormick's departure, including ship's artist Augustus Earle and Edward H. Hellyer, FitzRoy's personal clerk, who drowned on the Falkland Islands in March 1833 while pursuing a unique species of duck that he wanted to preserve and send back to naturalists in England. Benjamin Bynoe, assistant surgeon, and McCormick's replacement, bagged mammals, birds, and other animals and plants regularly, accompanying Darwin on many inland expeditions. But only Darwin filled the ecological-social niche FitzRoy needed, and his favor found the Cambridge-trained gentleman of leisure. The *Beagle* captain soon sealed their camaraderie with a gift to Henslow's young "geologist": a copy of Charles Lyell's new *Principles of Geology*, volume 1, which FitzRoy had already read. In retrospect, it was the most influential gift Darwin ever received.

Lyell's book codified the geological assertions made decades earlier by James Hutton, a Scottish acquaintance of Charles's grandfather, Erasmus, that Earth was very, very old, formed not by a snap of God's fingers on Day Three, but by internal heat pushing up and out followed by eons of erosion pulling in and down. No matter the geological formation, it formed through the same gradual processes that had been at work for all of Earth's history. Earthquakes and volcanoes, as violent as they appeared, were only minor blips in the imperceptible up and down of the "rock cycle." By extension, no global catastrophe, like Noah's flood, say, would account for the arrangement of fossils embedded in rocks. All Earth's features, even once biological ones like tree ferns and dinosaurs, became part of the geologic story very, very gradually. Water and wind sculpted grand canyons, isolated islands, and sky-scraping mountains imperceptibly slowly by uplift and erosion, said Lyell.

Presumably FitzRoy found Lyell's account important enough to foist it on wide-eyed Charles Darwin, who would have to stuff the tome, along with his other possessions for their two-year journey, in the 10' × 11' poop-cabin alongside two other companions, the large table for maps and charts, and hundreds of other books in the ship's library. Darwin had mostly planned on taking books on natural history. Henslow gifted him the seven-volume translation of Alexander von Humboldt's *Personal Narrative of Travels to the Equinoctial Regions of the New Continent*, the most celebrated account of a European scientist's exploration of South America. Darwin also intended to devote his time to studying "French, Spanish, Mathematics & a little Classics, perhaps not more than Greek Testament on Sundays." He was hoping to get a lot of reading done.

The two-year journey turned into three, then four, then almost five.

A year after departing England, at the height of the Tierra del Fuegian summer in January 1833, they successfully sailed up Tekenika Bay on the east coast of Hoste Island at the bottom tip of South America to return the Fuegians to "Woollÿa," Jemmy Button/Orundellico's ancestral home. The crew constructed three "wigwams" and planted a large vegetable garden for the

Fuegians and for missionary Richard Matthews, who sailed with the *Beagle* and befriended the Fuegians for over a year on ship. Matthews settled into his wigwam. FitzRoy, Darwin, and the others traveled west to continue surveying.

A few days later, a shocking sight. FitzRoy witnessed familiar looking fine English cloth adorning a strange native woman – cloth last seen on Fuegia Basket/Yokcushlu. Frantically, they returned east. There they found Matthews, unharmed but shaken. It had only been one week, yet Matthews had already been robbed, roughed up, and threatened with stoning. The vegetable garden stood in tatters. Only Jemmy/Orundellico remained, and he seemed socially isolated, only able to communicate with his brothers. After a quick discussion, Matthews abandoned his Christianizing mission until a more opportune time. A longer stay, he firmly believed, would have resulted in his death at the hands of unfamiliar Fuegians drawn to his cabin by the rumors of strange visitors bearing useful gifts, easily procured.

Upon revisiting Woollȳa over a year later in March 1834, FitzRoy found only Jemmy/Orundellico, a handful of his family, and the three huts remaining. Though happy enough, Jemmy/Orundellico appeared destitute, having been robbed by York Minster/Elleparu in collaboration with Fuegia/Yokcushlu and other members of a neighboring tribe. Though dispirited, FitzRoy nevertheless held onto hope that "a ship-wrecked seaman may hereafter receive help and kind treatment from Jemmy Button's children; prompted, as they can hardly fail to be, by the traditions they will have heard of men of other lands; and by an idea, however faint, of their duty to God as well as their neighbour." Years later, these hopes brought Rev. Matthews and others back to (re)establish the South American Mission near that spot.

Darwin spent much of the *Beagle* journey ashore in South America, a blessing, given how seasick he got. And, even when he wasn't with FitzRoy, he wasn't alone either. Teenaged fiddler and poop-cabin attendant Syms Covington (1816–61) proved a loyal servant for Darwin on land and sea, even if Darwin found him disagreeably odd at first. Covington helped with procuring fresh food and potable water, digging, shooting, preserving (perhaps according to the taxidermy skills Darwin learned from Edmonstone in Edinburgh), shipping specimens to Henslow, copying manuscripts, and most of the other functions that would eventually lead to Darwin's naturalist fame back in

England. They excavated fossil bones from the beach at Bahía Blanca in September 1832. They traveled with gauchos across Patagonia. Covington transcribed work on coral that became Darwin's first major publication. They survived fevers and cold and heat and earthquakes and gales. FitzRoy let Darwin hire Covington as a personal manservant at £60 a year (around £9,000 today) in 1833, but then agreed to keep paying half to feed and supply Covington. He continued to serve the Darwins through the 1830s, in London after debarking from the *Beagle* in 1836, even at The Mount in Shropshire, until Covington moved to Australia to pursue a non-servant's career and raise a family after Charles's marriage.

Darwin and FitzRoy got along for the most part, despite FitzRoy's tempestuous perfectionism. And the quarrels they did have Darwin courteously remembered as merely the effect of close quarters in the *Beagle* for so many years upon end. They had only one truly memorable fight, resolved the same day it occurred. That fight, in Darwin's recollection many decades later, was over slavery. Or at least this is how some modern scholars have parsed Darwin's words in his autobiography:

> [E]arly in the voyage at Bahia, in Brazil, [FitzRoy] defended and praised slavery, which I abominated, and told me that he had just visited a great slave-owner, who had called up many of his slaves and asked them whether they were happy, and whether they wished to be free, and all answered "No." I then asked him, perhaps with a sneer, whether he thought that the answer of slaves in the presence of their master was worth anything? This made him excessively angry, and he said that as I doubted his word we could not live any longer together.

FitzRoy recanted that outburst almost immediately. And note that FitzRoy's anger smoldered around Darwin's *manner* or the fact that he questioned the captain at all in his own rooms, not the principle of the argument. (Indeed, the fact that Darwin didn't fully *believe* FitzRoy caused another dispute between them before the *Beagle* left port in 1831.) Furthermore, FitzRoy would ascend to the governorship of New Zealand in 1843, but then he would be recalled only two years later because of his fierce defense of the native Māori *against* white settlers. Still, popular accounts both pit this episode as racist hothead FitzRoy versus abolition-minded Darwin and mark it as a central theme of their relationship throughout the 1831–36 voyage. At best, that's an exaggeration.

Their other major conflict, as it is often told and retold, occurred a few years later, on the Galápagos Islands, which poke out of the Pacific on the opposite side of South America over 600 miles west of Ecuador. This conflict arose from FitzRoy's commitment to Biblical literalism, including the associated Young-Earth Creationism – or at least this is how this myth goes. I'll address Darwin's religious proclivities in a later chapter. First, I want to turn to misconceptions over what happened on those crucial islands and what it meant for evolutionary theory.

2 The Truth About Atolls

Charles Darwin spent nearly the whole of his writing career attempting to convince his colleagues, the general public, and, by extension, you and me, that change occurs gradually. Tiny slivers of difference accumulate over time like grains of sand in a vast hourglass. Change happens, in other words. It's painfully slow, but it's inevitable. By implication, two organisms that look different enough to us to be classified as separate species share, many tens of thousands or even millions of generations back, the same ancestors. (Inbreeding means we don't even need to go back quite that many generations to demonstrate overlap, but you get the point.) But change that gradual means, as Darwin himself well recognized, that looking for "missing links" would be a pretty silly errand. Differences between one generation and the next look to our eyes just like common variation. It's one grain falling from the top of the hourglass to the bottom. You can't perceive the change. You would have to go back in time to find the very first individuals who possessed a particular trait – bat-like wings, say, or human-ish hands – and then, turning to their parents, you would see something almost identical. Change is slow, minute, fragment by almost-invisible fragment.

Weirdly, we don't treat *ideas* this way. We look, instead, for inventors, innovators, those people who saw something new in their heads and said, "Ah ha!" We hang on to stories of Steve Jobs at a whiteboard at Apple. Or Thomas Edison flicking on and off lightbulbs. Or Galileo peering at Jupiter's moons through a telescope. Or Archimedes slipping into a bathtub and watching the water level rise. "Eureka!" – we love that transformational moment, each exclamation point in the long human story. When retelling stories of ideas, we discount that *one small step for man*, because we so badly want that *one giant leap for mankind* that we think marks scientific change. But does that leap actually happen, really?

It didn't for Charles Darwin. Modern biologists might claim that Darwin became the famous scientist who we know him to be today on the Galápagos Islands. The truth is more complicated. There was no eureka moment for Darwin riding giant tortoises on the Galápagos. He didn't pluck finches out of the air, shipping them back to London as proof of his evolutionary ideas. He didn't think "evolution" on Culpepper (now Darwin) Island or see the diversity of marine iguanas as key to his theory while strolling to the British prison colony on Charles (now Floreana) Island. He didn't have anything so robust as a *theory* in the 35 days he spent in the archipelago in 1835. Yet the story we have collectively told ourselves since then has elevated Galápagos to a world-changing "eureka!" moment.

Descent, with Some Modification

He didn't do these things because he was still committed to the natural theological view that he carried with him as a recent Cambridge graduate. But that biographical factoid hardly matters. "Evolution" – by which we just mean the idea that groups of organisms transform over generations into other different sorts of organisms in a process known as descent with modification – was already an idea hatched in the heads of many men and women at the time. Even several of those who would become staunch opponents of Darwin's views in later years were *evolutionists*.

How could this be? In the simplest sense, already illustrated in the previous chapter, evolution in the broad sense is a very old idea. Charles Darwin did not invent, discover, or otherwise formulate the first theory of evolution. Precisely as he would say about the change from one kind of organism to another, the change from one evolutionary idea to the one introduced by Darwin would prove to be incremental, built from bits and pieces of lots of other individual ideas held by women and men in many cultures, speaking many languages, of many faiths. Some of those ideas were even labeled "evolution" before Darwin. Helpfully, he even told us whose bits and pieces he borrowed in the later editions of *On the Origin of Species*, just in case we were tempted to make him out to be an isolated genius – which, of course, we do anyway.

I suppose it's natural to think that he did invent/discover evolution. As many of my students tell me year after year, Darwin's name is about the only name they've ever heard attached to evolution; Darwinism the only scientific explanation for it. Sure, a few will note, "And there was this French guy who monkeyed around with giraffes." But even when I gently prompt, "Jean-Baptiste Lamarck...?" they squint and half-nod, skeptically.

The truth is, very many people were thinking about evolution while Darwin was gingerly stepping off the whitewashed boat onto the shores of San Cristobal Island in September 1835. It's just that Charles Darwin was not among them, as he scrambled on all-fours across the lava-rock shores punctuated with cactus. Instead, he was struck by the barren landscape. "We landed upon black, dismal-looking heaps of broken lava, forming a shore fit for Pandemonium [Hell's capital]," moaned Captain FitzRoy, echoing sentiments held by all the HMS *Beagle* voyagers. Thick silence greeted them, interrupted only by water against rock and the unsettling skittering of uncountable chitinous legs as crabs scrambled out of their way. (Six years later, American sailor Herman Melville would unknowingly repeat FitzRoy's bleak imagery upon encountering the Galápagos; years later the experience formed the scenic undercurrent to *Moby Dick*.) Even more than the land itself, Darwin found the marine iguanas, among other life teeming across the shores, "hideous." What an inauspicious waystation from which to launch the next stage of their circumnavigation, all the way across the vast, empty Pacific Ocean.

Darwin, with servant Syms Covington (Figure 2.1), FitzRoy, and others continued observing and collecting. Not every island presented itself like a small volcanic chunk of nightmare. Some featured squawking birds. Though by now Darwin had honed into a crackshot, shooting them proved largely unnecessary. He could virtually swat birds down with his hat. Iguanas he repeatedly threw into the ocean by their tails only for them to scramble up the banks again, perhaps more frightened of sharks than of a lanky, brown-bearded Englishman with a shotgun. The already-famous ancient tortoises lumbered along, paying them little attention. No animal showed any particular fear, fear that Darwin and his companions had assumed was natural.

Figure 2.1   Syms Covington, Darwin's personal servant and research assistant on the *Beagle* and for some months afterwards.

Today, we find the adaptation of plants and animals to these remote and largely inhospitable equatorial volcanic hotspots far from the mainland intriguing. It's noticing these adaptations, slightly different from island to island, we assume now, that gave Darwin that *one giant leap for mankind* insight that had him pondering evolution all the way across the unending Pacific, around Australia, through the tempestuous Indian Ocean, down and around Africa, then dispiritingly across the Atlantic to the east coast of South America again – because the Admiralty didn't like some of FitzRoy's measurements taken years earlier – and then finally, finally, finally, finally up to England's green shores in 1836. But no. In order to see transmutation, to think "evolution," 27-year-old Darwin needed to scramble on all-fours up the rock of centuries of scientific thought to stand, as fellow countryman Isaac Newton once put it, on the shoulders of giants.

The First Evolutionists

Aristotle saw it through the scales of fish and the slime of sea cucumbers on Isle Lesbos alongside his friend, Theophrastus, millennia earlier. Indian, Chinese, Persian, Egyptian, and Arabic scholars from Samarkand to Baghdad to Alexandria whispered it, too – species aren't stable; all life shares the same parents. But something happened in France in the 1600s and 1700s to jumpstart the modern search for hereditary resemblances, not just within families, but across otherwise seemingly unrelated plants and animals.

In his world-altering *Discourse on the Method* (1637), Réne Descartes scandalously imagined God spinning up a Universe parallel to our own like candyfloss (cotton candy) from practically indistinguishable particles and setting it loose to eddy into stars and continents, then iguanas and humans – a world "wholly like to one of ours." A century later, in the 1750s, Benoit de Maillet's anonymously published conversations with a so-called Indian philosopher suggesting that marine fossils embedded in rock cliffs high above the Mediterranean Ocean basin demonstrated Earth had hosted pre-human life for millions if not billions of years. In his contemporaneous *Essay on Cosmology* (1750), Pierre Louis Maupertuis envisioned vital particles combining and recombining to make many different forms of life. But only ones best adapted to their environments would persist to the present. Environmental pressures further altered those vital particles, which, when recombined from two parents in the next generation, led to still more possibility for adaptation – a concept called "pangenesis" that Charles Darwin himself echoed over a century later, in 1868. Swiss naturalist Charles Bonnet coined "evolution" as a biological concept in works on insects and plants from the 1740s to the 1770s, stressing that life would tend to progress in an unbroken chain from plants to invertebrates to higher animals, eventually humans. Each geological catastrophe would prove merely a "reset" for life to evolve still further.

With these figures in the immediate background, French polymath Georges-Louis Leclerc, Comte de Buffon (1707–88), bewigged monarchical director of the Royal Garden in the closing days of France's pre-Revolutionary splendor, set out to collect all known natural history. Over his comprehensive, many-volume *Natural History*, Buffon cast doubt on Bonnet's optimistic view of evolution. *Change* didn't necessarily mean *progress*, Buffon corrected. Instead, Buffon found evidence of de-evolution, regressing, descending from larger, superior-looking forms in Africa, Asia, and Europe to the strange, smaller, weaker inhabitants of North and South America and Australia. Adaptation to environmental conditions governed all life, except for the occasional "sport" or unexpected deviation, as Maupertuis indicated earlier. And even though Earth didn't seem to be quite as old as de Maillet speculated, Buffon believed there would be plenty of time for vital forces to take basically any general organic form to fit nearly all environmental conditions – fish to other fish, reptiles to other reptiles, and so on.

Descent with modification seemed readily apparent to Buffon's Anglo-American contacts Benjamin Franklin and Thomas Jefferson. Other scholars looked for even more evidence inside organisms themselves. To German contemporaries of Buffon and Franklin and Jefferson, namely Johann Friedrich Meckel and Carl Friedrich Kielmeyer, changes in embryos from more primitive forms to more advanced ones during gestation signaled a similar shift in the history of life itself; they called it "recapitulation."

Human change over time drew attention in the late 1700s and early 1800s from skull-measuring luminaries Johann Friedrich Blumenbach at Göttingen, John and William Hunter in London, and Charles White in Manchester. All four saw suspicious similarities between humans and non-human primates suggesting not just similar functional design ideas in the mind of the Designer but similar ancestry. And, as we've already seen, British physician Erasmus Darwin acquainted himself well enough with these concepts to popularize them in *Zoonomia* and *The Temple of Nature*, among other writings. No self-respecting Anglo-American or Continental naturalist in the Napoleonic Era could have avoided these evolutionary ideas.

But the most comprehensive evolutionary vision in the early nineteenth century came from a former French army lieutenant turned worm-expert, Jean-Baptiste Pierre Antoine de Monet, Chevalier de Lamarck (1744–1829). We just call him "Lamarck." After a career studying botany and insects on the one hand, the impact of fluid on the environment on the other, Lamarck wrote a series of books to complement that of his illustrious mentor, Buffon. Buffon cataloged; Lamarck explained. Buffon speculated that nebula cooled into stars and planets and that great epochs of different sorts of creatures followed one another over the course of Earth's history. Lamarck showed how fluids moved through the air in his *Research into the Causes of Principal Physical Facts*, published in 1794, and his series of *Meteorologies*, published between 1800 and 1810; then how fluids carved mountains and valleys in his *Hydrogeology*, published in 1802, and even shaped the living things that populated the land and air in *Zoological Philosophy*, published in 1809. From an extensive study of "invertebrates" – a term he coined, along with "biology" – Lamarck became convinced that species names were artificial designations. No firm line could be drawn around any group of animals even if there appeared to be clear distinctions between, say, insects and fish and amphibians and reptiles.

Instead, Lamarck concluded that the ladder of development from simple to complex organisms could be seen as a grand timeline (Figure 2.2). Life had been stretching, evolving, transmutating over eons. The operating substance of life, nervous fluid sparked by electricity, flowed through all animals and set them on a course of development. (His contemporary Mary Shelley borrowed from this view of the nature of life itself to ground her gothic horror masterpiece, *Frankenstein*.) Over the lifetime of each organism, this nervous fluid balanced the organism's needs against environmental pressures, strengthening certain organs in order to gain nourishment or perform other biological tasks, while other organs not used as often diminished, even disappeared – cave fish with no eyes, for example. Those acquired responses and the accompanying organs strengthened or weakened over the lifecycle of the organism to keep it alive and functioning.

Then the crucial biological material passed down from generation to generation – the *inheritance of acquired characteristics* that was already widely believed before Lamarck – such that children and grandchildren might share commonalities with, but specialize beyond, the confines of their parents and grandparents in response to the pressure to stay alive. One could easily imagine a planarian (flatworm), for instance, accumulating more nervous fluid in its head, translating into more sensitive eyespots to see better in order to capture prey. Or the nervous fluid inside a snail growing its foot just a bit to enable it to travel faster. Any tiny transmutation would not only ensure the continued survival of the line but alter it from its ancestors. After thousands or millions of generations, descendants would look almost nothing like their progenitors. And because electricity would spontaneously generate life and propel it up the evolutionary ladder, Lamarck's process replaced the simplest living things to fill any gap left by a transmutating set of organisms. If one species ascended to the next rung (owing to descent with modification) a less advanced species would be developing as well, in effect filling in behind on the vacated rung. At

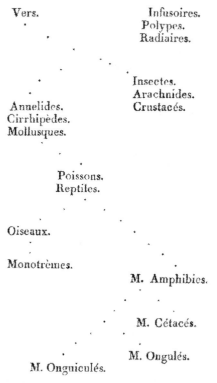

TABLEAU

Servant à montrer l'origine des différens animaux.

Vers.   Infusoires.
                                   Polypes.
                                   Radiaires.

                                   Insectes.
                                   Arachnides.
Annelides.   Crustacés.
Cirrhipèdes.
Mollusques.

                        Poissons.
                        Reptiles.

Oiseaux.

Monotrèmes.
                              M. Amphibies.

                        M. Cétacés.

                        M. Ongulés.
        M. Onguiculés.

Figure 2.2   Lamarck's evolutionary "tree" with worms evolving into insects, etc. From *Philosophie Zoologique* (1809).

the very bottom rung, life itself would be spontaneously generated to begin the upward climb. There were parallel chains of life from phytoplankton to whale, worm to human. Almost infinitely many parallel chains, never dropping a link, never leaving a gap.

It is easy for us to mock Lamarck's ideas. British figures known to and admired by Darwin, such as the world-renown geologist Charles Lyell, did attempt to explain away his transmutationism. But Lamarck's vision found plenty of scientific support on the Continent. For instance, in the wake of Lamarck's *Zoological Philosophy*, French embryologist Antoine Étienne Serres (1786–1868) dredged up the earlier recapitulation work of Meckel and Kielmeyer and fashioned the "Meckel–Serres Law" in the 1820s. Look at the development of the animal embryo, said Serres. See how it goes from undifferentiated goo to a more and more specialized organism with fins and scales or feathers and beaks or hair and hands? In those earlier stages, there is a great deal in common between the embryos of all types of animals. The differentiation only happens later in development and, in humans and apes, very late indeed. Interestingly, Serres went on, we see this very pattern echoed in the geological record: simple to complex, undifferentiated to very specialized. Perhaps, rather than environmental change impacting only the adult organism, environmental inputs redirect the development of embryos directly. This might seem like a silly idea when we observe only advanced mammals who spend the majority of their most fragile stages of development in a protected womb. But what about birds in eggs? What about butterflies that go through an egg stage, a caterpillar stage and a chrysalid stage before reproductive age – exposed to the rigors of the environment the whole time? And there are organisms with even more dramatic developmental trajectories. What if evolution worked by bumping developing organisms just a tiny bit in new directions, such that over time they could occupy different habitats, take on different appearances, interact with different predators and prey? Serres's amendment to Lamarckian evolution proved quite attractive.

These evolutionary ideas filtered into Edinburgh through, among other sources, the controversial anonymous geology article published by Robert Jameson in the introductory issue of the *Edinburgh New Philosophical Journal* in 1826 that I mentioned earlier. A more complete statement of the empirical backing for evolutionary theory had not appeared in English since Erasmus Darwin. It signals that a not-insignificant number of scientists in Edinburgh (some historians call them the "Edinburgh transformists") were willing to countenance evolutionary views – precisely when Charles Darwin was a student there.

Conspicuously anonymous (the only completely anonymous article among the 63 published in that issue), the author of "Observations on the nature and importance of geology" boldly claimed that, despite apparent boundaries between modern species, "various forms have been evolved from a primitive model ... species have arisen from an original generic form." Indeed, the anonymous author praised the "notions of the progressive formation of the organic world" by "meritorious" French evolutionist Jean-Baptiste Lamarck. According to the author's reading of Lamarck, "an aggregation process of animal elements" spontaneously generated two primeval forms: the most basic "infusory animals" (not yet named bacteria) and the "simplest worms." All other animals, the author claimed, "by the operation of external circumstances, are evolved from these in a double series, and in a gradual manner." The entire process could be seen in the geologic column: "We, in fact, meet with the more perfect classes of animals only in the more recent beds of rocks, and the most perfect, those closely allied to our own species, only in the most recent ... human remains are found only in alluvial soil...." What's more, the anonymous author used these geological observations to attack the entire notion of the stability of species, genera, and the rest of taxonomy: "The distinction of species is undoubtedly one of the foundations of natural history, and her character is the propagation of similar forms. But are these forms as immutable as some distinguished naturalists maintain...?" The author didn't think so. The evidence that species evolve could be plainly seen on the farm: "do not our domestic animals and our cultivated or artificial plants prove the contrary?" Breeders alter animals. In nature, random alterations in climate would mean alterations in food sources, which would require organisms to shift. And these changes necessitated many others, even in "relations" with other members of the species – the species would have to gradually adapt on all fronts or go extinct.

Why couldn't we see this change happening, then? Why did species *appear* to be stable? The author asserted that geology gave the answer here, too. Given how infrequently "petrifaction" (i.e., fossilization) happened, what geologists could see of this evolutionary process in rock could only be fleeting snapshots, "monuments concealed in the bosom of the earth." So, the author speculated, "many fossil species to which no originals can be found may not be extinct, but have gradually passed into others." All this has been taking place over eons of time so vast that "the duration even of the human race ... the few thousand years to which the [ancient Egyptian] mummy refers" would count as an insignificant whisper of time. Evolution erased its own tracks, leaving only the faintest markers of its imperceptibly gradual adaptive trail.

The Long Evolutionary Vision

What's the point of going down this winding (and very incomplete) path of Darwin's intellectual ancestors? I want to emphasize that very many features of Charles Darwin's thesis as disclosed in *On the Origin of Species*, eventually published in 1859, appear in one form or another in the writings of scientists, physicians, and scholars decades earlier. Let me spell them out:

(a) Earth is very old;
(b) Humans breed animals and plants into many different forms, sometimes dramatically unlike their wild selves – meaning organisms are "plastic";
(c) Forms similar to those familiar to us, but still different enough from present-day forms, exist as fossils in rock;
(d) Fossils of forms more similar to our own are in the most recent rock, and those the most different in the oldest rock;
(e) Embryological development echoes, at least abstractly, the same pattern we see geologically, with simple forms expanding and elaborating into more complex forms over time;
(f) Given (a) through (e), we can infer that all, or most, modern living things descended with modification from earlier ones;
(g) Gaps in the fossil record are merely the reflection of how unlikely it is to make a fossil at all; the embryological record shows the smooth transition of one form into another.

But how? It is the question of *how* evolution happened that provided the battleground. But not *if* it happened. Scientifically, the realization *that evolution had occurred* seemed less and less debatable over the course of the nineteenth century.

Erasmus Darwin, Jean-Baptiste Lamarck, Antoine Étienne Serres, and so many others I didn't mention in the above account simply assumed descent with modification was a *vital property* – just a built-in trait of living things to grow and change. *How*?! How was obvious, according to these early-nineteenth-century evolutionists. Growth and development could be seen in the way vines curled toward the sun, wrapping their way up trellises like slow snakes. And in the way blobs of yolk silently wriggled their way into the shape of feathers and beaks inside a chicken's egg. And in the way practically invisible pond scum could one day pierce the surface of the water, unfurling wings and a sharp mosquito proboscis to whirr near an observant scientist's ear. And in the way a bulbous fish-like tadpole could grow legs, could eat those mosquitos with a ridiculously long tongue that grew from . . . *where* exactly inside the tiny gelatinous frog's egg? Growth from seemingly nothing – from the invisible infusoria that naturalists assumed populated land, sea, river, lake, and air itself. Development from undifferentiated forms of embryonic slime into the most complex of creatures with hard beaks and bones and hair and teeth and feathers and blubber and whiskers and claws. Life cycles wherein worm-like insects who chewed their food for sustenance could transform into moths who drank from flowers exclusively through mouth-straws. Change from simple to complex, invisible to right there, alive and swimming and hopping and flying. It was all there to be seen by the naked eye. We just had to be observant enough, to really look.

Why should we think that, with all this sliding from one thing into another that life did all the time already, that sliding even further over immense amounts of time was impossible? Living things change. That is literally what it means to be alive. Given more time, they would change more, these naturalists insisted. Given eons, there was no limit to the sliding. God could have seeded the world in the beginning and let the sliding take its course.

Yes, but how?!?! only became a pertinent scientific question when someone with sufficient scientific stature planted his feet and yelled, "Stop!" That man was French comparative anatomist Georges Cuvier. And by the time Darwin scrambled up the lava rock shores of the Galápagos to toss iguanas and ride tortoises in 1835, Cuvier and his followers had muted the nineteenth-century exuberance over transmutationism trumpeted by the likes of Erasmus Darwin and Lamarck and the Edinburgh transmutationists.

The Man Yelling "Stop!"

Lamarck's chief opponent turned out to be a coworker. Georges Cuvier (1769–1832; Figure 2.3) was younger than Lamarck and without Lamarck's military bona fides. Yet Cuvier established himself as the best comparative anatomist alive and one of the best-known scientists around the globe in the early nineteenth century. He claimed to be able to reconstruct the relevant details of an entire organism from a single tooth or bone. He could do this, Cuvier explained, because of two interrelated iron laws of living things. First, all parts of an organism had to work together for its survival. Teeth, eyes, muscles, fur, stomach, liver, gall bladder, even toenails all played distinct, irreplaceable roles. Every single part *correlated* to make a fully adapted whole animal, snout to tail. Change a small thing – the shape of a flipper, say, or the number and style of teeth – and a complete reconfiguration would be required. And, secondly, the "conditions of existence" mandated this strict

Figure 2.3   Georges Cuvier at the height of scientific prominence.

correlation of parts. Only the well-adapted survived to reproduce. Why can't cheetahs retract their claws like other cats? Why are their backs hinged to allow maximum stride? Why do their tear-ducts carve such aggressive channels away from their eyes? Because, Cuvier insisted, the cheetah's conditions of existence mandate that too much deviation from a life spent nabbing antelopes at breakneck speeds over dry grasslands means certain death.

Indeed, death has occurred. Mass death. Full extinctions, even, insisted Cuvier in the 1810s and 1820s. Catastrophic floods, volcanoes, and earthquakes throwing up the sea floor to create mountains (Cuvier had seen seashells in high places, too). These geologic events unfortunately meant that, over the vast history of Earth, millions of organisms have suddenly found themselves occupying new conditions of existence without any ability to adapt, given how rigidly the coordination of parts makes each animal and plant. So, mass death. Extinction.

Life finds a way, though. How? Cuvier suggested that other already existing organisms must have rushed in to repopulate areas opened by these extinction-level catastrophes. Jaguars and tigers and cheetahs must have always been there, ready to rush in when saber-toothed *Smilodon* went extinct, in other words. But that's not what the fossils say. Fossil evidence, already in great abundance by the 1820s, suggested modern plants and animals didn't exist in any recognizable form in the past. So, these new occupants, the ones that we recognize today, where did they come from?

Completely rejecting both colleague Lamarck's transmutation and a simplistic interpretation of Genesis, Cuvier left open the possibility for a "God of the gaps" explanation: maybe new creation events occasionally replenished regions devastated by catastrophes. Maybe each epoch of time really signaled a new series of creations, perhaps not across the entire planet, but in certain zones. Sure, it wasn't *Genesis*, but it was still creationism. How else could we account for the fact that life always found a way? Transmutation? Don't be silly: change means death.

Cuvier's "catastrophism" – massive extinction events, visible in the geological record, followed by either repopulation from existing organisms or new creations – found adherents in Britain, including prominent naturalists, who were also churchmen: William Buckland at Oxford and Darwin's own mentors, Adam Sedgwick and John S. Henslow at Cambridge. Charles Darwin certainly learned about Cuvier's arguments through them.

Captain FitzRoy had likely heard these arguments as well. That makes his gift to Darwin of Lyell's *Principles of Geology* – a book that fundamentally under-cut the geology portion of Cuvier's catastrophism by mandating only gradual change – an unmistakable gesture. (Again, histories of Darwin's *Beagle* voyage are apt to paint FitzRoy as a close-minded Biblical literalist; his pushing of Lyell's theories on Darwin, and not the reverse, put the lie to this.)

Lamarck died in 1829 and, just before the *Beagle* journeyed around South America in the early 1830s, Cuvier pounced, attempting to dismantle Lamarck's transmutationism. His direct criticisms fell on two basically unknown French mollusk hunters. Indirectly, however, Cuvier took aim at another distinguished colleague and acolyte of Lamarck, zoologist Etienne Geoffroy Saint-Hilaire (1772–1844), who upheld his own version of transmutationism. Sparring for years, the two officially locked horns before the French Academy of Science in early 1830, while Darwin was still at Cambridge. Geoffroy preserved their debate in *Principes de Philosophie Zoologique* (1830), an account of their opening salvoes with a title suspiciously like Lamarck's most famous book. European intellectuals, from German luminary Johann Wolfgang von Goethe to Cambridge's original philosopher of science William Whewell to Britain's own Cuvier impersonator, the masterful anat-omist of dinosaurs and extinct flightless birds Richard Owen, read and remarked upon the Geoffroy versus Cuvier debate. Through them, the issues perhaps filtered to Darwin, too.

In print, Geoffroy continued wrangling until Cuvier's death in the spring of 1832, while Darwin was on the east coast of South America. Before he died, Cuvier claimed victory. Adaptation was too important; the struggle for exist-ence too overpowering; even small mutations too disruptive. The kind of change visible in the geologic record could only be possible through major extinction-level catastrophes. Transmutationism sounded nice, and indeed the transmutationists appeared to find "homologies," or common parts between animals related by descent – the way the bones in a bat wing and a human hand and a seal flipper all seem to be the same bones in the same arrangement, just different in size. But they only *looked* the same, said Cuvier.

Functionally, they were nothing alike. And in an environment to which the organism must-must-*must* be adapted, the function of survival and reproduction governed all else. No one had demonstrated how small changes could be translated into the actual structure of the organism to be passed on to the next generation. No amount of fluid dynamics inside a worm or an insect could create new organs to help them develop new habits to allow them to evolve. The struggle for existence would snuff out that misfit creature.

Train in Vain

So, we shouldn't be surprised that Darwin uttered nothing like an evolutionary theory while swatting birds on four of the sixteen main Galápagos Islands – then named Charles, James, Chatham, and the largest, Albemarle. He did *not* note that some of those birds had different-sized beaks depending upon what particular island they lived. That realization, for which Darwin is often admired today, took place later and only with the expertise of ornithologists such as John Gould, largely unrecognized by the public today.

Instead, Darwin, FitzRoy, surgeon Bynoe, and several other *Beagle* men ascended the heights of a small subset of the islands, capturing specimens of various kinds. As a team, they found copious numbers of reptiles and birds but comparatively few insects – this fact Darwin-the-beetle-collector found disappointing and a little mysterious. Their specimen labeling wasn't always the most careful, so they might have missed some other mysteries. But the Galápagos Island collection proved a nice comparison to the Falkland Islands they had visited before crossing the Straits of Magellan, and, before that, Cape Verde and the islands off the African coast. In the next year, they would visit the islands of Tahiti, New Zealand, Tasmania, Cocos (Keeling), various atolls across the Indian Ocean, and Mauritius before landing at Napoleon's final prison in the South Atlantic, St. Helena, in 1836. Darwin became intimately familiar with the kind of white sand beaches, palm trees, and tropical sunsets for which people today pay good money to visit on holiday. Actually, to be fair, over 1835–36, Darwin rapidly became a world-class expert on coral reef formation.

Of course, that's not why he's well-known now. Now he is known for his theorizing about plant and animal evolution. But search Darwin's account of the *Beagle* journey, first published in 1839, and you find page after page of *geology* after his visit to the Galápagos, not *biology*. The puzzles that might have sparked him to think of natural selection float through the account of the *Beagle* voyage unexplored. Read only his "*Voyage of the* Beagle," in other words, and you won't come away believing that Charles Darwin's thoughts veered in the same direction as his evolutionary grandfather.

Even his extensive 1845 second edition of the *Beagle* journal, revised after Darwin had already penned over 200 pages spelling out his hunches about descent with modification, offers only hints that he was interested in animal change over vast geological time. Certainly, even avid readers of his account – Alfred Russel Wallace was one – found Darwin stubbornly unwilling to speculate about the causes behind his and FitzRoy's collections of fossils and birds and shells and flora on Charles and Albemarle.

Nevertheless, the puzzles were there. Wallace felt them. If adaptation to environment governed all else – as Cuvier insisted, as Henslow and the other natural theologians believed, as even the transmutationist Lamarck assumed – then why were the collection of plants and animals on all these island groups so different? Weren't volcanic islands scattered across the great oceans of the globe nearly the same sorts of climes? Some of them were clearly tropical atolls, built in equatorial waters from immense coral colonies. Shouldn't the living things growing there be perfectly adapted to bright sun, ocean breezes, and relatively little rainfall? Darwin assured his readers that the landmasses did appear to be almost carbon copies of each other no matter what ocean they appeared in. Why, then, did atoll occupants seem more similar to the nearest continental flora and fauna, even if those continents possessed vastly different conditions of existence? Why were those Galápagos birds that he could almost pluck off a tree with his hands (he initially called them "blackbirds" and "gross-bills," and only found a portion of the finches later attributed to him) so similar to the ones Darwin bagged with his shotgun 1,000 miles to the southeast in Chile? Even more mysterious, why weren't they *identical* to the ones on the next island over?

Answering these questions would require a team of experts, few of whom gained the notoriety Darwin did for collecting the birds and fossils that generated the questions. We'll meet some of them in the next chapter.

So, no, it was not in the Galápagos that "Darwin became Darwinian." Nor in Sydney, Montevideo, Bahía, Hobart, or Cape Town. It took going back to London, being ensconced again in a network of scientists, some of whom disagreed with the transmutationism coming out of France, once championed by Darwin's grandfather, now by Robert Grant. And it took conversations with guests at his brother Ras's parties – people who had read more than Charles.

This misconception about Darwin's transformation on the *Beagle* has spawned other ones, perhaps even more pernicious, small myths that I feel compelled to swat down like a Galápagos bird. Here's a brightly colored one: retellers of Darwin legends depict an evangelical FitzRoy arguing that God created each species of Galápagos finch specifically for each island and a scientific Darwin arguing that all were related through common descent. Supporters of this myth even have a few quotes they can pull: for instance, "All the small birds that live on these lava-covered islands have short beaks, very thick at the base, like that of a bull-finch.... In picking up insects, or seeds which lie on hard iron-like lava, the superiority of such beaks over delicate ones, cannot, I think, be doubted," and "[E]very animal varies more or less in outward form and appearance." To the Darwin-evolutionist/FitzRoy-creationist crowd, these seem like excellent evidence.

Except that these quotes are from *Captain FitzRoy*, not Charles Darwin.

FitzRoy noted adaptation. He noted variation. True, FitzRoy mentioned these beaks "appear to be one of those admirable provisions of Infinite Wisdom by which each created thing is adapted to the place for which it was intended." But this milquetoast reference could hardly be taken for staunch Biblical literalism.

I think we can glean something much more interesting from FitzRoy's observations. FitzRoy made his bird beak comment in the context of discussing the lack of fresh water on many of the equatorial islands. Stout bird beaks only *appear* to be for picking up food. In reality, they are still more useful for squeezing berries, cutting through tree pulp and leathery leaves to get to water. It's the absence of water, even more than food, that makes life so difficult on these islands, that makes it so difficult to understand how some foreign bird, plant, or, say, massive tortoise, could get established in the Galápagos in the first place.

For help with an explanation of the origin of species on the Galápagos, FitzRoy went to another account, namely that of James Colnett, captain of the HMS *Rattler* on a June 1794 investigation of the archipelago. Captain Colnett noted that some flora and fauna on the Galápagos appeared to be emigrants from South America, though some birds appeared even more similar to New Zealand varieties. The cacti, which supplied a good bit of the water for the island inhabitants, could be found, Colnett believed, on other tropical islands in the Pacific and even the Atlantic. As FitzRoy deduced after reflecting on Colnett's experience alongside his own on both *Beagle* trips, many of the inhabitants of these desert islands flew or floated in. Given the massive die-offs of birds, tortoises, turtles, and snakes that Captain Colnett reported occurred during the dry season, the struggle for existence must be severe. Probably some "refreshing" of new organisms from the not-that-nearby South American continent must happen to keep the Galápagos populated.

But FitzRoy expressed another theory, one that again shows he was not locked into any defense of Genesis versus an evolutionary Darwin. How did the massive tortoises, who moved slowly and swam little, if at all, make it across open ocean? The question troubled Charles Darwin. FitzRoy provided a simple answer; he was, in fact, a bit annoyed at "the short-sightedness of some men, who think themselves keener in discrimination than most others," who claimed tortoises couldn't migrate, so must have been originally created on and for the South American archipelago. FitzRoy scoffed at the creationists: "there is no other animal in the whole creation so easily caught, so portable, requiring so little food for a long period, and at the same time so likely to have been carried, for food, by the aborigines who probably visited the Galápagos Islands on their *balsas*, or in large double canoes, long before Columbus. . . ." Darwin, in fact, easily nabbed a small tortoise, not for science, but to keep as a pet on the *Beagle*. And the *Beagle* crew took several large adults as a food supply across their long Pacific leg of the journey. As for the other flightless inhabitants? FitzRoy noted a very warm current drifting south from the Panamanian isthmus and a cold one blasting north from Peru and Chile. In combination, they created a "gulf stream," a watery gyre spinning around the islands, that could pull objects from the continent at a comparatively high rate of speed to deposit on the nearly barren shores. One might imagine, Darwin would write years later in his *Journal of Researches* (now published as *The*

Voyage of the Beagle) that all the birds, all the cacti, all the flies and iguanas and flowers and coconuts came from a small number of initial visitors carried by wind and waves to these uninviting shores. By his own account, FitzRoy agreed – no defense of Special Creation here.

Once a small food chain established itself, via a few hardy cacti and insects, perhaps, sea birds would set up rookeries, just as on other distant "guano" islands. Then land birds, perhaps blown in during storms. Meanwhile, as Captain Colnett noted nearly a half-century earlier, branches, roots, and seeds still fully capable of germination and growth washed up continually on the beaches. It was just one of those barely noticeable, gradual things that happened on every beach in the world, much like Lyell proposed with rocks. Even though thirst perpetually led to die-offs in the dry season, some drifted-in water-retaining plant life stuck and flourished, drawing up rainwater and dew. Animals learned to chip through the bark and suck the liquid from leaves. Others gathered around natural basins in the ground, awaiting rain. Eventually Aboriginal sailors from the western shores of South America carrying drought-resistant giant tortoises followed those cold currents racing north, deposited eggs or even breeding pairs on the islands. And those gnarled, wizened reptiles, with memories built from lifetimes two or three times longer than our own, would ever after trudge up the Galápagos mountain paths to wallow around happily in the same mud pits they had when hatchlings an age ago, to reenact an ancient, almost sacred dance in the struggle for existence on these dry equatorial islands.

So, both Darwin and FitzRoy acknowledged the gradual process by which organisms sprinkle out from one geography to another. FitzRoy, more force-fully than Darwin, highlighted the probable role of prehistoric humans in that redistribution process. At the time, *neither* envisioned the islands of the Pacific as evidence in favor of Erasmus Darwin's vision – fresh and new when Captain Colnett visited the Galápagos Islands – of common descent with modification, that "the whole is one family of one parent."

3 London Calling, 1836–1842

John Gould's father was a gardener. A very, very good one – good enough to be head of the Royal Gardens at Windsor. John apprenticed, too, becoming a gardener in his own right at Ripley Castle, Yorkshire, in 1825. As good as he was at flowers and trees, birds became young John Gould's true passion early in life. Like John Edmonstone, John Gould (1804–81) adopted Charles Waterton's preservation techniques that kept taxidermied bird feathers crisp and vibrant for decades (some still exist in museums today), and he began to employ the technique to make extra cash. He sold preserved birds and their eggs to fancy Eton schoolboys near his father's work. His collecting side-hustle soon landed him a professional post: curator and preserver of the new Zoological Society of London. They paid him £100 a year, a respectable sum for an uneducated son of a gardener, though not enough to make him Charles Darwin's social equal (Darwin initially received a £400 annual allowance from his father plus £10,000 as a wedding present).

Whatever his social class, the *Beagle* birds, some 450 of them, showed up on Gould's desk on 4 January 1837.

By then, Gould was the best-known taxidermist in all of England. When Windsor Castle needed new stuffed animal decor, they came to their old gardener's son. The East India Company came to him. So did the British Museum. King George IV's giraffe died in 1829, a few months before its overstuffed owner; naturally the crown put Gould in charge of stuffing it. And beginning in 1833, Gould's friendship with Irish MP Nicholas Aylward Vigors, an enthusiastic ornithologist, culminated with Gould being named superintendent of the ornithology section of the Zoological Society. From there, Gould befriended another accomplished ornithologist, Sir William Jardine. Vigors and Jardine helped Gould organize and identify hundreds of bird specimens sent from far-off central Asia by the East India Company in the late 1820s. Gould's new wife, Elizabeth Coxen, related to a London taxidermist herself and an accomplished visual artist, created the stunning lithographs for the resulting book, *A Century of Birds from the Himalaya Mountains* (1830). Aside from launching into the new, commercially successful enterprise of bird illustration, the exercise made the Goulds experts at spotting even minute similarities and differences between birds.

On 11 January, less than one week after receiving the birds from Darwin, FitzRoy, and the other *Beagle* voyagers, John Gould and his ornithological network identified the Galápagos Archipelago bird specimens.

Unfortunately, Darwin labeled his birds only haphazardly. One couldn't tell from which island each bird derived. Syms Covington and Captain FitzRoy fared somewhat better with their quarry. Yet, the labeling mess turned out to be a happy accident in the end. Not predisposed to connect geography to classification, the Zoological Society ornithologists found that the collection Darwin identified as containing "blackbirds," "mockingbirds," and "grossfinches" was in truth twelve species of finches. They were fairly plain-looking for tropical birds, mostly a dull brown color. Many of them had never been seen anywhere but on that distant collection of volcanic islands (Figure 3.1).

Had John Gould followed the strict theories of Georges Cuvier that the internal coordination of animal parts, especially those related to the *conditions of existence* for the organism, governed everything else – and thus must be inviolable on pains of, eventually, extinction – he likely would have ignored important information he deduced from these small brown birds. At minimum, he wouldn't have passed it along to his peers in the Zoological Society. Nor even to Darwin. Because in that grouping of birds that Gould insisted were all different species of finches, it was their beaks that differed. *Beaks*. The eating parts. The parts most important for a bird's existence. How could the same kind of bird have different mouths? This had nothing to do with Special Creation; this was a scientific question. How could the *mouth* change without causing such a drastic rearrangement of all the other coordinated parts

Figure 3.1 A few of "Darwin's" finches collected by the group of HMS *Beagle* explorers. Later ornithologists attached the labels displayed.

of the bird? Wouldn't the feathers change long before the beaks? But the feathers were very close in color and arrangement, much closer than the beaks. What kind of mystery did Gould's, er, Darwin's finches expose?

Career Opportunities

Darwin didn't learn any of this until he moved to London months later, in the spring of 1837. He had survived his globetrotting *Beagle* adventure. Anxious for the familiar, he sprinted to The Mount after landing in October 1836, accompanied still by servant Syms Covington. He popped around to his brother Ras's place in London. Then he moved back to Christ's College and fell in with his old professors, Henslow and Sedgwick, again. Henslow handled many of Darwin's specimens returned by packet ship over the voyage, and it was Henslow who pushed Darwin into the limelight through his fossil discoveries in South America. Now it was time for Darwin to begin the sorting and labeling process. Henslow also built the bridges to the Who's Who of British science, scions like comparative anatomist Richard Owen (the man who coined "dinosaur") and Darwin's geological inspiration, Charles Lyell. Sedgwick found Darwin a comfortable three-story townhouse to let at 22

Fitzwilliam St., just across from old Peterhouse College. He had few peers, seeing as how multiple years of classes had come and gone. Still, it felt nice returning to that comfortable old Cambridge life. For the better part of four months, Darwin sorted specimens during the day, farming them off to other experts if possible, and, in the evenings, drank a fair bit again, placing bets against students and other fellows (against the rules) and languishing too late in the dining hall (also off-limits; he paid fines for such actions in bottles of quality port). He loved the extended college glory days – too much, in fact. By 1837, he felt the clash of adult life, heard London calling. His brother had a place prepared for him. Perhaps he would return to Cambridge as a professor, but not yet.

It's questionable whether this was a wise move for bachelor Charles. Brother Ras never made MD. He completed his Bachelor of Medicine at Christ's. But his father found him too sickly to practice or to continue to train in vain (this was a family trait). So, Robert Darwin "pensioned" his 26-year-old son. With dad's cash, Ras moved to trendy Soho, bringing his new "air-bed," and began living it up as a bachelor. In the years since Charles departed on the *Beagle*, Ras tried charming Frances "Fanny" Mackintosh Wedgwood, wife of his cousin, the local police magistrate Hensleigh Wedgwood. Despite (or perhaps because) scandal seemingly always lurked nearby, the three created a hub of young, radical literati – a Soho Circle. For a time, edgy, nearly deaf author Harriet Martineau (1802–76) pried Ras away from Fanny; Charles thought Harriet would make a good sister-in-law. But Martineau continually scolded Ras for his spoiled lackadaisicalness. Everyone knew even Charles worked harder than his older brother, and that was saying something.

Neither of them approached the work ethic of Martineau, whose hearing impairment drove her to write more and more frequently and forcefully throughout her life. It's quite likely that she set the table for, if not directly inspired, Charles Darwin's second pivotal read (the first being Lyell's *Principles of Geology*): Rev. Thomas Malthus's *Essay on Population*, published in 1798.

Over the 1830s, Martineau catapulted into intellectual prominence with her prodigious speed in writing economic explainers in favor of breaking down the old aristocracy (Figure 3.2). In *Illustrations of Political Economy*, published

Harriet Martineau.

Figure 3.2 Harriet Martineau exposited Malthusian theory to popular audiences and at Ras Darwin's dinner parties.

between 1832 and 1834), a series of short stories with economic lessons embedded and explicated at the end, she demonstrated the disparate impact of British economic policy on the lives of actual Britons. *Poor Laws and Paupers Illustrated*, published between 1833 and 1834, and *Illustrations of Taxation*, published in 1834, editorialized on the economics of Jeremy Bentham, James Mill, Adam Smith, and especially Thomas Malthus and David Ricardo. These works placed her among the best-known apologists for "classical" economics, as it would later be called. She argued aggressively for the Ricardian position that the old English Poor Laws should be torn up.

While Darwin was in South America in the mid-1830s, Martineau visited the United States. Her *Society in America*, published in 1837, excoriated King Cotton and the slaving system in the USA (ironically, her family money derived from cloth manufacturing in Norwich). Her book served as a contemporary corrective to the better-known Alexis de Tocqueville's *Democracy in America*, published just two years earlier. Unlike de Tocqueville, Martineau pronounced that the United States had betrayed its foundational democratic premises by, among other things, allowing enslavers to capture whole states' economies and political systems. Hypocritically, the enslavers and those who profited from slavery still gathered in their Baptist churches every Sunday, praising their white, cotton-robed Jesus. Also unlike de Tocqueville, she lambasted the lack of women's control over their own destinies in the States. American women, Martineau found, hovered barely above the enslaved in social standing and control over their own bodies. Breeding and rearing young seemed to be the only thing American men permitted women to do – and nearly all women went along with it, later reinforcing the norm even more forcefully than men. Not surprisingly, given these statements, credible threats of lynching followed Martineau all the way back to England.

Back in London, Ras Darwin followed her like a wealthy puppy "morning, noon, and night." But, once he had met the couple after returning from the *Beagle* voyage, Charles assured his sisters (and through them, his disapproving father) that nothing untoward could be happening. Darwin was keeping an eye on them from his own new bachelor pad on Great Marlborough St., Soho, on the same busy block as his brother and the Wedgwood cousins. From Charles's perspective, Martineau couldn't really be a marriage prospect, being physically impaired as well as radical, a believer in women's equality, abolition, and assumably other taboo topics. She talked too much and self-promoted shamelessly, gossiped Charles. She was bossy; she worked his poor older brother like he was her "n****r" (yes, he used that expletive, though he said it was Erasmus's expression; given her experience in the USA, Martineau would not have been amused). Plus, as he mentioned to family in November 1836, Charles found her "plain" (DCP 321). His assurances (if we want to call them that) turned out to be right, in a sense: Ras and Harriet never married.

Nevertheless, around Martineau, London intellectuals and poseurs hovered like bees. A little shy, perhaps, but explorer Charles fit right into their hive. They threw great parties at which Martineau held court, even among the Who's Who. Her *Illustrations of Political Economy* blew up; booksellers couldn't keep copies on the shelves. In them she taught Smithian division of labor, railed against mercantilism and monopoly, chastised idleness (with her finger seemingly pointed squarely at the wealthy Darwin brothers), pitched suggestions to solve the Catholic troubles in Ireland, mocked the era's obsession with celebrity and silly fads. And, in one especial downer-of-a-tale, entitled "Weal & Woe in Garveloch," Martineau explicitly championed the social economics of Malthusianism.

Too many mouths to feed – it was a mathematical certainty, explained Martineau channeling Malthus. The government, wanting to address growing hunger, might offer some little financial relief to buy a little more food. And it would seem a Christian kindness. But then the poor would just go and have more children – again, a mathematical certainty that population out-reproduced resources – and everyone would be back in the same miserable place, with not enough food. Eventually, there wouldn't be enough land to house everyone, enough farms to feed them. War, disease, every bad thing, would result. These, she called "positive checks" ("positive" in that they stopped population growth, they were actual, active stops, not because they were good). Given that the goals of human existence were, according to Martineau, "virtue and happiness," it would be better to employ "preventative checks" – birth control and/or abortion, in other words. That would be the innovation that would get humanity out of this Malthusian trap (Figure 3.3).

Famously, Charles Darwin would repeat much of this argument after reading Malthus "for amusement" in September and October 1838. If he heard this exposition of Malthus originally from Martineau, he never credited her. (But then Darwin didn't credit any other woman for any "discoveries," for that matter.) It's not because he didn't like her; on the contrary, he found her "wonderful."

Perhaps he was too star-struck in those circles to recall where he first heard ideas. Harriet and Ras also invited over geologist (and Darwin's idol) Charles Lyell.

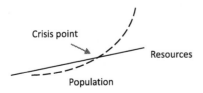

Figure 3.3   Malthusian theory (simplified).

Charles Babbage, the mathematician and proto-computer scientist, too. Self-important essayist Thomas Carlyle, who had just completed a ground-breaking three-volume history of the French Revolution, came over regularly to grouse. So did parson-naturalist Leonard Jenyns, the man originally chosen to accompany FitzRoy on the *Beagle*. Frankly, Charles seemed to be the least witty, least well-read of the entire Soho Circle. On evening after evening, the group danced between politics (Whigs, mostly, though Martineau showed much too much sympathy for working men), economics (Ricardo, mostly), foreign relations (pro-abolition, though Carlyle, especially, proclaimed his own disgusting affinity for permanent Black servitude), religion (German higher criticism throwing doubt on all Biblical miracles and perhaps the Incarnation and Resurrection), and science (Lyell's old Earth gradualism). Actual dancing often ensued.

After some months, in May 1838, Charles earned a private audience with Martineau. They talked intensely, author-to-author. Darwin wrote too slowly; she offered tips as a formidable writer, even generously volunteering to observe Charles writing in situ. But it's hard to know if he took her intellect seriously: he reported to sister Susan merely excellent flirting with his flirtatious brother's significant other. Four months later, on 28 September 1838, Darwin opened Malthus's *On Population* "for amusement." An instant flicker of recognition: too many mouths to feed; too few resources; conflict, disease, and, unless some new innovation came about, misery and death. He had been bathed in this talk at party after party with some of the most influential intellectuals in London for almost two years. But only now could Darwin see it as "a theory by which to work." This radical, disabled woman who brought her own version of *laissez-faire* economics to the English *nouveau riche* through writing and witty dinner *soirés* – who never earned the approval of the Darwin brothers' father, Robert, nor finally convinced Erasmus to give up his "air-bed"-enhanced Soho bachelor lifestyle – likewise never made an entry in Charles Darwin's intellectual self-accounting.

The Right Profile

Boats never became Darwin's thing. After the *Beagle*, he never made an ocean voyage again. But maybe, given his *Beagle* voyage, preceded by the Firth of Forth sea sponge work alongside Robert Grant, the ocean got into his bones somehow anyway. His second major publication turned out to be about those coral atolls they visited all across the Pacific and Indian Oceans, *The Structure and Distribution of Coral Reefs*, published in 1842. In midlife, Darwin returned to study barnacles for eight headache-filled, seemingly interminable years, culminating in four volumes in the 1850s. And when sick – which he seemed to be all the time after returning to England – he visited doctors espousing the "water cure," hoping to uncork his nervous system by hyper-hydration. Water, water everywhere.

No wonder, then, that he didn't immediately publish work on those Galápagos Island birds when he received the report by John Gould's Zoological Society ornithologists that his seeming disparate collection of birds comprised strikingly related finches. Compiling the list of where he and FitzRoy and Covington and other *Beagle* members collected those birds turned out to be far too difficult to trace them to individual islands with different conditions of existence. He suspected they were tightly correlated to climates on each different island, but he didn't know for certain. The collecting and labeling were just too haphazard. But it didn't matter, really. He was already becoming more well known for rocks and fossils as a Lyell protégé (Figure 3.4).

Darwin's speculations about the atolls, sent to Henslow while Darwin was still onboard the *Beagle*, created some buzz. Upon his return home, Darwin found himself already ushered into British scientific society's good graces. He delivered similar papers to Lyell's Geological Society of London during his years at 36 Great Marlborough St., in Soho, from 1837 to 1839. Soon he became Geological Society secretary. Those years would be his only "normal," in-person interaction with the scientific world. He would slowly recede, preferring prodigious correspondence to scientific meetings.

He married his one-year-older first-cousin Emma Wedgwood on 29 January 1839; they moved to 12 Upper Gower St., in Bloomsbury (a bit further out of the Soho Circle's gravitational pull), and he launched into even more prestigious circles as a Fellow of the Royal Society. FitzRoy's massive report on the second voyage of the *Beagle* featured Darwin's *Journal of Researches*, published in 1839, as the third volume. Far from positing trans-mutation like his grandfather Erasmus, Darwin wondered from what "center of

Figure 3.4 Charles Lyell, from attorney to geologist, Darwin influencer, and opponent of transmutation.

creation" the current occupants of the archipelago originally hailed. (FitzRoy, now Governor of New Zealand, already conjectured on this point: the west coast of South America.) The creationist language was intentional. All the radicalism of the Soho Circle hadn't quite seeped in. When illustrious anatomist Richard Owen's report on Darwin's prehistoric South American fossils revealed spooky similarities with modern-day South American rodents, everyone in London and Oxbridge who mattered viewed Charles Darwin as an up-and-comer, a real *geologist*, very much the fully orthodox, fully respectable gentleman naturalist he had aspired to be in order to join Captain FitzRoy in 1831.

Through it all, he promoted Lyellian British gradualism, not godless Lamarckian French materialism. Lyell himself came out forcefully against Lamarck's (and grandfather Erasmus's) transmutationism in the second volume of the *Principles of Geology*. The book was already in its sixth edition, and given his notes in the margins, Darwin read the more recent edition even more carefully than the one he had with him on the *Beagle*.

The problem with French/Scottish transmutation, said Lyell, *isn't* that naturalists should believe in Special Creation (contrary to the propaganda of both today's evolutionary biologists and Creationists, already by the 1830s, naturalists and churchmen thought it demeaning to God to make constant interventions in creation necessary; would you rather believe God is an *incompetent* designer who changes His mind and keeps needing to fix His junk handiwork?). The problem was *evidence*. We witness dog and cat breeding. Breeders can radically change dogs' and cats' bodies. But they can't make dogs or cats grow new organs, which Lamarck said would happen with constant pressure to change the conditions of existence. Look at 6,000-year-old Egyptian mummified dogs and cats – where it counts, they're exactly like today's dogs and cats. They haven't so much as changed their number of teeth or their whiskers or their skull shapes. True, an old Earth means there's lots of time for change. But Lyell couldn't recall a single piece of actual hard scientific evidence the Lamarckians could bring forth in support of their position. Yes, one could see resemblances between living things and fossilized things. But that just indicates that there used to be more things around, and then some of them died off. Evidence like that demonstrates extinction, sure, but not transmutation, per se.

In the end, Lyell insisted the geological, zoological, and botanical evidence went so far as to undercut species *stasis*, because organisms can vary enormously, the way that dogs and cats and livestock breeds do. But no evidence touched species *stability*. In other words, variation could be enormous, but it always fluctuated around a core, almost Platonic, idea of an organism. You know a cat or a horse or a rose or a beech tree when you see one. That's what we mean, ultimately, by genera: the type that gets reproduced – sure, with significant variations in dimension or coloring or behavior. And maybe we could go so far as to say that multiple species could originate from one genus depending on environmental conditions, the way that different species of buffalo live in Africa, Asia, and the Americas. But the extreme sliding of one thing into another championed by Lamarck and grandfather Erasmus has never been shown. None of the French transmutationists could ever produce evidence to show that loping wolves, scurrying foxes, jogging jackals, and snorting pugs have all mutated one from another, for instance. Ridiculous. House cats from leopards and cheetahs? A stretch. And humans from non-humans? That was right out.

For all intents and purposes, Charles Darwin echoed his mentor during his London years – or at least avoided confrontation in print. In the background, though, Darwin had something else fermenting.

"I think. . ."

Fast forward to 9 March 2022. A pink gift bag mysteriously appeared in front of an archivist's office at Cambridge University Library. The card attached read:

Librarian,

Happy Easter,

X

Inside, two small boxes, each containing a small, worn, red leatherbound notebook carefully wrapped in plastic. These approximately 4-inch-wide by 6-inch-long handheld notebooks turned out to be two of the most famous objects in the life sciences: notebooks Darwin jotted in during those years in London in the late 1830s. Today, we call them the "Transmutation Notebooks." They had been missing ever since being sent for digital photographic preservation, when that technology was in its adolescence, over 20 years earlier. The celebratory noise of their recovery rippled around the globe in 2022 – just when the world needed a bit of good news.

"Zoonomia," Charles Darwin titled the clasped notebook he opened around July 1837, later labeled "Notebook B." And though four decades later in his autobiography he would downplay any deeper connection, the reference to his grandfather's work should be glaring. The fact that his old transmutationist mentor Robert Grant was just then rabble-rousing, no longer from Edinburgh, but from University College London, only a few blocks away from the town home Darwin would occupy with Emma after their marriage, also suggests a firmer connection than Darwin wanted to recall years later. Even with all the caveats and qualifiers in place, these notebooks contain the musings of a young man wandering into the same dense thicket of speculations his grandfather, French naturalists, radicals like Grant, and some of Ras's dinner guests had explored.

"Two kinds of generation," Darwin began Notebook B, "the coeval kind, all individuals absolutely similar; for instance fruit trees, probably polypi, gemmiparous propagation. bisection of Planariæ. &c &c. – The ordinary kind which is a longer process, the new individual passing through several stages (typical, or shortened repetition of what the original molecule has done)."

The second kind of generation – sexual reproduction launching into an embryological process that repeated the geological or evolutionary process – was where his grandfather found the most interesting material as well. "[A]ll warm-blooded animals have arisen from one living filament," grandfather Erasmus spouted, "which THE GREAT FIRST CAUSE endued with animality, with the power of acquiring new parts, . . . possessing the faculty of continuing to improve by its own inherent activity, and of delivering down those improvements by generation to its posterity, world without end!" Erasmus trumpeted this on page 505 of the first volume of his *Zoonomia*. We know Charles read it because he underlined it. Grandfather Erasmus envisioned a vital power shoving organisms from the inside, driving them into new and improved adaptations. Lamarck had said this, too, though Lamarck appealed largely to the same natural processes that allowed rivers to delve through rock until organisms formed "habits." Serres, too, emphasized that we could look to the embryological to see the evolutionary path trod by that "original molecule." Secretly – sheepishly, even – Charles Darwin was bringing the family business back from France, making descent with modification Darwinian again.

As he read more of his grandfather's work, which by then was poopooed by hard-nosed, celebrated scientists like Charles Lyell, but championed by other hard-nosed, socially marginalized scientists like Grant, Darwin's thoughts carved deeper into less explored parts of the transmutationist thicket. Down in there, he found a tree.

Lamarck's tree grew upside-down, a real tree of descent. Lamarck saw infusoria, tiny creatures growing into worms, insects, larger and more complex as they cascaded down, from amphibians to reptiles, pulled inevitably toward more complexity, herbivores to carnivores, four feet to two, tree monkeys to ground apes to (presumably) big-headed humans. But Cuvier and his followers, even Lyell, shut that speculation down, especially the closer it got to threatening the vaunted place of humanity. No one had shown – really illustrated with hard evidence – how so much as a canine or a wing or a swim bladder could be lost without dooming the whole line of organisms to extinction.

A few dozen pages into his "Zoonomia" notebook, after a scribbled "I think...," Darwin drew his tree right-side-up with four twigs labeled "A" through "D" (Figure 3.5). Unlike Lamarck, he didn't start trying to account for all of animal life. "D", he labeled a left-wise twig off of a larger branch full of twigs; "B" at the top twig of the top branch; "C" just a bit further down from "B" and across from "D"; "A" jutting off one twig on one of three branches on

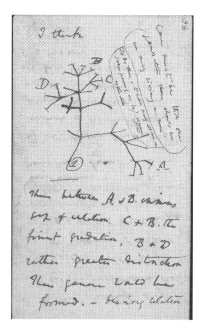

Figure 3.5 Biology's most iconic scribble in Darwin's hand-held (and, until recently, lost) "Zoonomia" Notebook "B."

a crazy far-off limb in the right corner. Had it been a living tree, "A" would have scraped the ground far from the trunk and just as far from "B."

> Thus between A. & B. immens [sic] gap of relation. C & B. the finest gradation, B & D rather greater distinction Thus genera would be formed. – bearing relation to ancient types. – with several extinct forms, for if each species an ancient(1) [sic] is capable of making, 13 recent forms. – Twelve of the contemporarys [sic] must have left no offspring at all, so as to keep number of species constant. (36–37)

Imagining extinction plucked other species off other twigs in his tree diagram, naturalists would find "B" and "C" closely related species. "D" could be a more distant cousin species. And – imagining these were Galápagos finches, for example – "A" a truly odd bird.

He clearly believed he depicted *species* here, all species of the same genera, in fact. Given that number, "twelve of the contemporarys [sic]," which coincided with what John Gould made of Darwin's plus FitzRoy's plus Covington's Galápagos Archipelago finches, it's reasonable to assume this is what Darwin had in mind, or something near to it. He had gone to a novel place, connecting new natural evidence with an older hypothesis of adaptative radiation. But it was still a relatively conservative conjecture. So, he must have shocked himself when his own avalanche of speculation gained speed.

> If we thus go very far back to look to the source of the Mammalian type of organization; it is extremely improbable that any of the successors of his relatives shall now exist, [. . .] Hence if this is true, that the greater the groups the greater the.gaps (or solutions of continuous structure) between them – for instance there would be great gap between birds & mammalia, Still greater between Vertebrate and Articulata. still greater between animals & Plants. But yet besides affinities from three elements, from the infinite variations, & all coming from one stock & obeying one law, they may approach, – some birds may approach animals, & some of the vertebrata invertebrates. (40–44)

Indeed, just as grandfather Erasmus had said, all could be descended from one parental stock! Extinction would lop off some of the connecting branches and twigs, leaving only a few living examples in species, perhaps even genera or families, leaving big gaps between them. Related organisms could fill the gaps left by an extinct species and then adapt to a new environment. This was another big realization, melding some of FitzRoy's observations with Lyell's gradualism and some French (gasp!) transmutationism.

Perhaps most importantly for Darwin, these private ideas presented a crucial departure from what he originally gleaned from William Paley's natural theology as a Cambridge undergraduate and from Henslow and Sedgwick, channeling Georges Cuvier: adaptation needn't be so tight. Organisms might be able to thrive even in novel environments to which they couldn't have been adapted initially. Owls could have potentially carried rodents from a different South American island to the barren volcanic rocks of the Galápagos. There, the escaped rodents flourished. Seeds and coconuts could have washed ashore, riding those currents north up the continent as described by FitzRoy, digging roots into gaps in the rock. Birds, blown in by a storm, finding a dry landscape more friendly to iguanas. Tortoises carted in for centuries as the slowly walking meal pack of the ocean explorer. It seemed strange to imagine, yet there they all were, American colonists on the equator, hundreds of miles from their original homes. Adapted to a different environment, still showing markings of that former existence. Life found a way.

This was what all the fuss was about when the "B Notebook," the "Zoonomia" notebook, went missing in 2001, and why Easter 2022 proved to be such a happy occasion at Cambridge University Library. Those handheld, leather-bound notebooks contained the first grains of Darwin's Big Idea. In the summer of 1837, 22 years before his *Origin of Species*, before ever reading Malthus or publishing his *Beagle* material, without acknowledging his fore-runners, mentors, and interlocutors, and without seeing what those Galápagos birds really meant, Charles Darwin nevertheless offered in his uncertain ("I think...") scrawl this plausible – and more importantly *iconic* – depiction of evolution as a diverging, very branchy tree.

4 Darwin–Wallaceism

Meeting the "White Raja of Sarawak" in Singapore in 1853 had been a stroke of luck. Honestly, it could have been a major turning point in what had been an unlucky career so far for 30-year-old collector Alfred Russel Wallace (1823–1913) (Figure 4.1). But the steep, rocky, sweaty climb up Borneo's Mt. Serembu (also known as Bung Moan or Bukit Peninjau) in the last week of December 1855 wasn't exactly what Wallace expected. His eyeglasses fogged in the humidity. Bamboo taller than buildings crowded the narrow path. Near the top, the rainforest finally parted. But it revealed neither a temple nor some sort of massive colonial complex with all the trappings of empire worthy of a "raja." Instead, there leaned a modest, very un-colonial-ruler-like white cabin. When he saw it, Wallace literally called it "rude."

The White Raja himself, Sir James Brooke (1803–68), appeared in the doorway, greeting Wallace and his traveling companions with open arms, magnanimously, as always. This was what he was best at: the old Bengal Army man "possessed in a pre-eminent degree the art of making every one around him comfortable and happy." Wallace immediately strode up onto the tiny porch and located the rocking chair. He was right at home.

It's true that the White Raja's abode down below in Sarawak was more comfortable. Wallace left the rude white cabin to spend Christmas 1855 down there with Sir James, some European expats enjoying the White Raja's company, and several other companions, some British or Dutch, some Malay. That was colonial. In fact, that was his second Very Colonial Christmas in the company of the White Raja in his walled compound in Sarawak. Wallace left a lasting impression on the guests and on Sir James himself, insisting with

Figure 4.1   Alfred Russel Wallace in 1889.

a straight face that "our ugly neighbors" the orangutans in the Borneo jungle were human ancestors. They enjoyed his explorer's tales, but that company was not where Wallace felt best. After Christmas 1855 ended, Wallace and his exploring companions left those finer accommodations and trudged back up Mt. Serembu to the plain white cabin. It became their base camp for the first four weeks of 1856, their collecting hub. They nabbed samples of everything they could. Insects, mostly. And, despite the fact the people were rumored to be headhunters, Wallace's crew boldly left the cabin to meet the local Dyak people who lived in tiny villages on the Serembu slopes.

The cabin turned out to be important for Wallace's scientific theory. Moths on Serembu presented a special kind of puzzle for him. Over one week in December 1855 plus the first half of January 1856, he brought in a huge haul: almost 1,400 moths. Mostly, he netted them by the handful here or there. It was really on just four very wet nights that Wallace and his companions had the most success, capturing roughly 200 per night. Funny enough, they saw almost no more moths anywhere in Borneo. The moth drought continued for weeks. Then months. Then years. Indeed, the White Raja's white shack would become the only place in the eight years he spent in the widely dispersed, diverse Malay Archipelago (now Indonesia) where he caught so many. Why could he catch them only here in this white clapboard cabin near the summit of a mountain and, even then, only on rainy nights? Were moths cut off from moving off this island, somehow? They had wings, but were they *stuck*?

But this was only one of the mysteries Wallace encountered on Borneo. There were the orangutans – or *mias*, as the Dyak called them (and Wallace, interestingly, followed the locals' customs). After shooting 16 at various locations deep in the rainforest and examining preserved skeletons of others, Wallace disputed the traditional classification of orangutans into three different species. The mystery was in their sexual dimorphism – the feature that can make male and female individuals so physically different from one another. He admitted that he'd seen far too few specimens to make definitive claims, but there appeared to be just one species, though modified dramatically (modern scientists agree). *What could that mean?* Wallace wondered.

And then there was the mystery of the Dyak themselves. Far from being hostile, they warmly welcomed Wallace into their villages circling Serembu. They seemed to earnestly enjoy the Europeans, unlike many of the Malay who Wallace had encountered elsewhere in Borneo. How were Dyaks related to the other people groups of Southeast Asia? Perhaps unsurprisingly, he took the next step of a British naturalist, wondering how all these *races* of people he encountered should be ranked against each other – a topic to which he would return as an expert ethnologist back in London. In the meantime, he noted something truly strange. Europeans who had encountered the Dyak regarded them as lowly savages, biologically more closely related to beasts, more exposed to the vagaries of the natural world, since culture hadn't yet lifted them up to the level of the English. The plentiful food in the area meant the Dyak should, like beasts, have large broods of children running around. Yet Wallace observed the opposite: Dyak seemed to have smaller families than even the civilized English. This was odd. According to the demographic theories of Rev. Thomas Malthus, which Wallace knew well, the reproductive rates of these closer-to-nature people should be much higher than whites. Malthus insisted that natural "checks" would be needed to keep population levels down, otherwise beasts, and the savage people like them, would reproduce without any cap on their numbers. But as Wallace sat and ate and explored with these supposedly more primitive Dyak people, he noted that they had regulated *themselves* into smaller families. The Dyak averaged only three or four children instead of five or more that civilized Europeans expected in their families. And it wasn't because of higher infant mortality. *What could that mean?* he wondered again.

The mysteries continued to pile up the longer he remained in the southeast Pacific archipelago. Some of the mysteries reflected those Wallace first saw when he traveled in the Amazon River basin alongside the great butterfly hunter Henry Bates years earlier. Now it had mounded into an *intercontinental* pile of mysteries.

Just the fact that so many similar mysteries appeared on opposite sides of the globe from one another indicated something crucial to Wallace. All those biogeographical explanations he had been taught since he was young – what every expert European naturalist just *knew* to be true – didn't accord very well with what Wallace saw with his own eyes in South America and now in the East Indies. Something was wrong with the science. Someone had to say something about it.

Wallace believed he was onto something big, a scientific theory that might overturn what all those naturalists said the world should be like. It took him until 1855 to write it all up. He penned a journal article at Sarawak which appeared in the September issue of *Annals and Magazine of Natural History* (since 1967 renamed the *Journal of Natural History*): "On the law which has regulated the introduction of new species." Why shouldn't the origin of species be governed by a regularly operating natural law? Wallace thought the regularity he observed both in Borneo and in the Amazon must be just the working out of that law. He framed it like this in 1855: "Every species has come into existence coincident both in space and time with a pre-existing closely allied species" (188). What that meant, if naturalists took it seriously, could overturn all our under-standing of life on Earth, Wallace believed.

Still, it's not this discovery (or its publication) for which we remember Alfred Russel Wallace. In its day, that 1855 paper received very little attention. It is still basically ignored today. We still talk about Wallace because of what he would write in a surprise 1858 letter mailed directly to Charles Darwin. This letter made Wallace, according to many people – even Wallace himself – the co-discoverer of the *process* of evolution: natural selection. And he based his theory, just like Darwin did, on the theory of Rev. Thomas Malthus.

But are their theories truly identical? If so, why don't we call them "Wallace–Darwinism" or "Darwin–Wallaceism"? In this chapter, we're really trying to answer three interrelated questions:

(1)   Did Wallace and Darwin discover the same scientific principle?
(2)   Is "Wallaceism" the same thing as Darwinism?
(3)   Why does Wallace receive substantially less recognition than Darwin?

The Contender

Organisms vary. That much is obvious. Wallace only proclaimed this obvious point in his 1855 "Introduction of new species" article because he thought deep time qualitatively altered what this obvious point meant. Given the vast amount of time geology shows us has occurred, those variations would accumulate and accumulate. And occasionally something would happen to make the variations more dramatic, larger.

No serious scientist disagreed with this part of his argument when Wallace made it in 1855. Lyell, still among the best-known scientists in England, had written that whole chapter in recent editions of *Principles of Geology* to deal with just this problem. Lyell knew that godless French scientists propped up their theories of transmutation on this point. By 1844, Lyell knew the enemy was inside the walls: that year some anonymous author from somewhere in Britain published *Vestiges of the Natural History of Creation* advocating evolution from top to bottom, stars to humans. It cited Lamarck. It cited the Meckel–Serres Law that "ontogeny" (meaning, the development of embryos) recapitulates "phylogeny" (meaning, the fossil record from simple to complex). Evolution's fingerprints appeared everywhere. Many, many people read *Vestiges*, including Wallace, who found the book convincing. But the transmutationists still had not produced convincing evidence. *Vestiges* was sloppy, as both Darwin and Wallace believed. Moreover, how could science keep humanity from being scooped up in that same evolutionary theory if science accepted it for sea sponges and moths and primroses and oaks and apes? In the end, this is what bothered Lyell the most – humans have to be set apart, he thought.

Wallace knew Lyell's objections almost as well as Darwin did. As another passionate observer and collector, however, he saw evidence that the dramatic, larger sorts of variations – the kind that would make a naturalist question whether two organisms belonged to the same sub-species anymore – happened quite a lot, really. If two groups became separated by geographic barriers from each other – by mountains, say (moth-haven Mt. Serembu might be one), or waterways like the ocean around the Malay Archipelago or the immense Amazon River – the two varieties of organisms would begin to diverge, looking and acting more and more different from one another. They would stop breeding with one another. Eventually, they would have changed so much that you couldn't even tell they had once been related. As Wallace signaled to the readers of his 1855 paper, so many confusing things in natural history would just make more sense if we could imagine groups of organisms altered over time, gradually slipping from one species to a new and different species. As Darwin's "bulldog" Thomas Huxley remarked after rereading the 1855 paper years later, Wallace had decisively laid down the scientific gauntlet: either disparate acts of creation were needed to pop each species into existence, or more recent species descended from former ones with modification along the way. No middle path existed between these explanations.

If this all sounds suspiciously similar to Darwin's *On the Origin of Species*, published just four years later, that's because in many ways it is. The two men shared a remarkable number of things in common; perhaps we shouldn't be surprised that a possible explanation of descent with modification occurred to both of them. Wallace grew up only one county south of Darwin near the Welsh border in the west of England. They both read Malthus, *Vestiges*, and the French transformists. Wallace read Darwin's account of the *Beagle* voyage before he left on his own trek in Southeast Asia (Figure 4.2). He very much appreciated it as an explorer's story, the same way Darwin

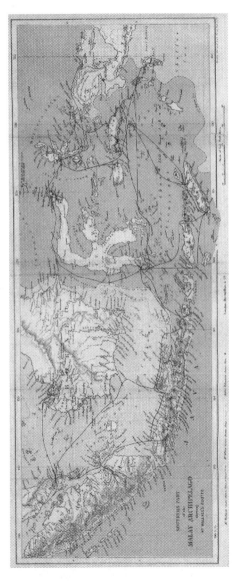

Figure 4.2   The line depicts Wallace's travels through the "Malay Archipelago," now Indonesia.

viewed the travels of Alexander von Humboldt and Charles Waterton a generation earlier.

But it was Darwin's 35 days spent in the Galápagos Islands two decades earlier that now jumped out at Wallace, given that he was on a different set of islands on the opposite side of the globe. There was something big hidden in the Galápagos Archipelago, but Darwin hadn't seen it, Wallace thought. (At this point, he had no idea Darwin had worked on his "theory" since that "Zoonomia" notebook in 1837, revealing it to only those he believed he could trust.)

Look again at the Galápagos, directed Wallace in his 1855 paper. As readers of Darwin's and FitzRoy's accounts of the second *Beagle* voyage well knew, they "contain little groups of plants and animals peculiar to themselves, but most nearly allied to those of South America. . . ." But why should that be the case, Wallace wondered. When reading Darwin's *Journal of Researches* (now in a revised 1845 edition – today we call that version of the book *The Voyage of the* Beagle), Wallace could find no explanation offered by the older naturalist, not even a "conjectural" one. Wallace boldly leapt in with one of his own:

> [The Galápagos] must have been first peopled, like other newly-formed islands, by the action of winds and currents, and at a period sufficiently remote to have had the original species die out, and the modified proto-types only remain. In the same way we can account for the separate islands having each their peculiar species, either on the supposition that the same original emigration peopled the whole of the islands with the same species from which differently modified prototypes were created, or that the islands were successively peopled from each other, but that new species have been created in each on the plan of the pre-existing ones (187).

No wonder those Galápagos mockingbirds and finches and tortoises and iguanas and cacti and so on looked similar but not exactly alike. As volcanic island species far from the mainland, organisms living on each Galápagos island certainly would demonstrate similar adaptations. But the real story was less about adaptation than about their shared *histories*.

After originally "peopling" the islands – seeds being blown in during a storm, say, or insect eggs hitchhiking on birds' feathers – each animal and plant group would have remained separated from each other and from their parent species in South America. Wallace speculated that over time each organism would "modify" from their "prototypes," becoming their own "peculiar species," built from the "plan of the pre-existing ones." Something similar must have happened in these South Pacific Malay islands as well. His thousands and thousands of collected species showed this.

If you think about it metaphorically, mused Wallace, the pattern of life on one of these islands must look something like a branching tree. In some cases, maybe even most, descendant species would simply replace parental species – he used the strange word "antitype" to mean parental species (that word might have led to some problems, as we'll see below). A diagram would reveal "several species in direct succession in a straight line." But now think about related species on separate islands. Those could only be represented by a "forked or many-branched line." Each species of mockingbird, say, would be "independently formed on the plan of a common antitype," in other words, an ancestor shared by both sets of birds. After a few cycles of species replacement, each island would have its own forked line of mockingbirds as descendants replaced ancestors, each generation just a bit different than the last. The collective history of mockingbirds across the dozens of islands would look still more branched. All mockingbirds in that region? Or even that continent, then – whew! How could you even diagram that? Wallace tried to illustrate that complexity: "in the actual state of nature it is almost impossible, the species being so numerous and the modifications of form and structure so varied, arising probably from the immense number of species which have served as antitypes for the existing species." The naturalist would end up drawing lines of descent "as intricate as the twigs of a gnarled oak or the vascular system of the human body." One can imagine Darwin's famous "I think. . ." tree appearing in Wallace's mind as well (actually, Darwin saw that very thing in Wallace's writing). Except in Wallace's version, the "I think. . ." is vast, interwoven, a spider's web, a vine-covered tree in the Borneo rainforest.

But the scientific world did not respond overwhelmingly to Wallace's claims about the origin of species. Even Darwin ignored it at first. After reading the 1855 paper, which he apparently would not have done without prompting from a Wallace ally, Calcutta naturalist Edward Blyth, Darwin scribbled "nothing very new" (DCP 1792, n1). But historians have pondered why that would have been Darwin's reaction. Perhaps Wallace's choice of terminology tripped him up. "It seems all creation with him," grumbled Darwin in the margins of the paper. And it's true, Wallace did use that term "creation" eight times in the 1855 paper when meaning "origination" or "appearance." Yet Wallace just as frequently insisted the present distribution of living things "derived by a *natural* process of *gradual* extinction and creation of species...." Wallace didn't mean Genesis, even if he never used "transmutation" or the still rarer term "evolution." Moreover, Darwin himself would close the *Origin of Species* by referencing the Creator *breathing* life – a much more literal scriptural allusion than anything Wallace used. But Wallace also deployed that strange word "antitype." Taken outside of a biological context, the term means "foreshadowing symbol" and was once a specialized term most used in theological discussions. When Darwin did get around to reading Wallace's 1855 article, he mentioned more positives than negatives, admitting in his marginal comments in December, "Explains Rudimentary organs on same idea (I sh$^{d.}$ state that put generation for creation & I quite agree)" (DCP 1792, n1). So, given that Wallace made no mention of a heavenly creator, we have to ask, was Darwin being deliberately obtuse about that term "created"?

We should consider the larger social context as well. Though his father had been a lawyer, Alfred Russel Wallace grew up relatively poor. He began work at a young age and took his last formal classes at age 14. When his father died, his mother, desperate for work, became a housekeeper. They shipped his unmarried sister, Fanny, to the United States to teach at Wesleyan Female College, deep in the slave-owning heart of Georgia. Wallace would never breathe the rarefied Oxbridge air of a Darwin. These facts placed Wallace as a relatively unknown *collector*, not an Oxbridge-trained *naturalist*. "But the mere collector is not and cannot be justly considered as a naturalist," stipulated John Barlow Burton, the eminent British entomologist.

The social distinctions can be seen from their first interaction in the summer of 1856, when Darwin asked Wallace to catch some jungle fowl. Darwin seemed to be looking for "one of the originals of the domestic breed of poultry" (DCP 2004, n4). It's quite possible that Darwin didn't even read Wallace's 1855 article until 1857, when he wrote to Wallace asking for more information about leopards, the domestication of poultry, and snails. It was in that 1857 letter that Darwin wrote to Wallace regarding the existence of his much larger project: "This summer will make the 20th year(!) since I opened my first-note-book, on the question how & in what way do species & varieties differ from each other," Darwin revealed. "I am now preparing my work for publication, but I find the subject so very large, that though I have written many chapters, I do not suppose I shall go to press for two years" (DCP 2086). But it's hard to say it was a shot across the bow, warning Wallace to stay out of Darwin's way. Indeed, it's probable Wallace read it in quite the opposite way, given that Darwin belatedly lauded Wallace's otherwise unacknowledged 1855 paper on the natural law originating species:

> I agree to the truth of almost every word of your paper; & I daresay that you will agree with me that it is very rare to find oneself agreeing pretty closely with any theoretical paper; for it is lamentable how each man draws his own different conclusions from the very same fact. (DCP 2086)

Wallace must have been greatly encouraged at Darwin's response. And besides, collecting thousands of beetles and moths and butterflies and every other kind of life to populate European museums without so much as a whisper of gratitude from the British naturalist community – it felt lonely. Any tiny crumb of acknowledgment, Wallace greedily gobbled up.

Despite the agonizing slowness of correspondence from the other side of the planet, Wallace still managed to inform Darwin in September 1857 that he, too, was working on a bigger project: "The mere statement & illustration of the theory in that paper is of course but preliminary to an attempt at a detailed proof of it, the plan of which I have arranged, & in part written..." (DCP 2145).

Darwin encouraged that, too. As he responded in December: "I am extremely glad to hear that you are attending to distribution in accordance with theoretical ideas. I am a firm believer, that without speculation there is no good & original observation" (DCP 2192). This accounted for the lack of response regarding Wallace's 1855 article as well. Most naturalists merely blindly collect without theorizing, Darwin comforted. (Curious statements coming from Darwin, who claimed elsewhere to be a firm believer that a good scientist accumulated facts *without* preconceived speculation.) And look, Darwin appeared to assure Wallace, there are people in your corner; you have influential allies. Blyth in Calcutta was the one who had introduced Darwin to Wallace's paper. That was a big deal. And Charles Lyell, one of the most important scientists in the entire world, had also spoken highly of Wallace's speculations. (Wallace must have felt his heart leap when reading that.) In any case, Wallace was also onto something important out there on those islands, Darwin insisted, and he should keep working at it: "Few travellers have [at]tended to such points as you are now at work on; & indeed the whole subject of distribution of animals is dreadfully behind that of Plants." And Darwin made sure to mention again the book he was writing: "My work, on which I have now been at work more or less for 20 years, will not fix or settle anything; but I hope it will aid by giving a large collection of facts with one definite end. ...." But he neglected to say exactly what it was that he was writing. Was it treading the same ground as Wallace? Was his book just about domesticated varieties and species? Was it just about plants? It certainly wasn't going to be about human beings: "You ask whether I shall discuss 'man'; – I think I shall avoid whole subject, as so surrounded with prejudices, though I fully admit that it is the highest & most interesting problem for the naturalist." And, whatever it was, Darwin was ploddingly slow at getting it out into the world. After 20 years (according to Darwin; but he hadn't actually published a word of it), and even without the major life hurdles Wallace had to leap over in order to publish anything, Darwin still ambled. "I get on very slowly," he apologized, "partly from ill-health, partly from being a very slow worker ... I have now been three whole months on one chapter on Hybridism!" (DCP 2192). To Wallace, Darwin must have seemed to let the "perfect" book be the enemy of the "good" or even the "finished."

Surprise!

Wallace's (in)famous letter to Charles Lyell laying out his evolutionary hypothesis arrived in Darwin's hands in June 1858. Wallace mailed it in March from Ternate in the Maluku Islands (the Moluccas or "Spice Islands" west of Papau New Guinea), where he had acquired a terrible bout of a tropical illness, probably malaria. He hoped Darwin would send on the enclosed essay, titled "On the Tendency of Varieties to Depart Indefinitely from the Original Type," to Charles Lyell. Darwin would be the surrogate.

Wallace presented his argument more directly than Darwin had yet voiced: (1) organisms spend most of their time procuring food and trying not to die; (2) all variations (e.g., color, hairiness, limb dimension) will result in either a shrinking population of those with the variation or a growing one; (3) if shrinking, they will eventually go extinct; if growing, they will eventually replace their parent species.

Despite their now years of earlier communication, when Darwin read the letter and enclosed essay, his heart skipped a beat. Wallace beat him to the punch. But it's hard to account for the surprise: Why exactly didn't Darwin see how close Wallace was to publishing something of his own, something that would "scoop" Darwin?

After reading Wallace's letter on 18 June 1858, he immediately wrote to Lyell. It was perhaps the hardest letter he'd ever written in his adult life.

My dear Lyell –

Some year or so ago, you recommended me to read a paper by Wallace in the <u>Annals</u>, which had interested you & as I was writing to him, I knew this would please him much, so I told him. He has to day sent me the enclosed & asked me to forward it to you. It seems to me well worth reading. Your words have come true with a vengeance that I shd. be forestalled. You said this when I explained to you here very briefly my views of "Natural Selection" depending on the Struggle for existence. – I never saw a more striking coincidence. if Wallace had my M.S. sketch written out in 1842 he could not have made a better short abstract! Even his terms now stand as Heads of my Chapters.

Please return me the M.S. which he does not say he wishes me to publish; but I shall of course at once write & offer to send to any Journal. So all my originality, whatever it may amount to, will be smashed. Though my Book, if it will ever have any value, will not be deteriorated; as all the labour consists in the application of the theory.

I hope you will approve of Wallace's sketch, that I may tell him what you say.

My dear Lyell / Yours most truly / C. Darwin (DCP 2285)

The whole month was difficult. One of his surviving daughters, Henrietta ("Etty"), came down with a case of diphtheria. Then the Darwin household became the site of a scarlet fever outbreak. Their toddler, Charles Waring Darwin, died of it on 23 June 1858, only a few days after Darwin's reading Wallace's letter. Later, members of the Darwin family revealed that they suspected Charles Waring was developmentally delayed. Emma was nearly 50, after all. But in those days, a malformed child might signify something sinister in the blood. With the niggling doubt left by the death of poor little Charles Waring, Darwin would, a decade later, prod one of his adult sons, George, to do some statistical investigation into the health challenges of children born to near-relatives – meaning his and Emma's first-cousin marriage, of course.

Why did Darwin write to Lyell after receiving Wallace's natural selection letter in June 1858? Hadn't Lyell opposed transmutation back in the 1830s? Indeed, he had. But Wallace asked for it, and Darwin, honorably, followed through. More than that, however, over their decades-long relationship, Darwin had burnished his zoological *bona fides* with his large, serious, four-volume study of barnacles. He was now Lyell's scientific peer. All along, in the background, Darwin chipped away at Lyell's resistance to transmutationism, perhaps even lending Lyell Darwin's own copies of Lamarck's *Zoological Philosophy*. Privately, Lyell's cracks were showing.

Even more importantly, beginning with an auspicious meeting at the prestigious Kew Botanical Gardens, Darwin constructed a web of prominent correspondents who were more sympathetic to transmutationism, including Joseph D. Hooker, the polar explorer and chief botanist at Kew (Figure 4.3), and Asa Gray, distinguished botanist at Harvard University in the United States. Amidst hundreds of other scientific correspondents, Darwin also found common cause with "young Turks" including insect and snail expert Thomas Vernon Wollaston (1822–78) and anatomist Thomas Henry Huxley (1825–95) (Figure 4.4). Friends like these buoyed Darwin when sparring with Lyell, maybe even wore down his own timidity about showing his cards.

On one rowdy weekend in April 1856, Darwin, Wollaston, Hooker, and Huxley flamboyantly scoffed at the idea of species altogether. Lyell heard of their coterie, from which he had been excluded despite the fact he had visited Darwin only a few days earlier. He wrote a friend, aghast, "I cannot easily see

Figure 4.3   Sir Joseph Dalton Hooker in 1851.

Figure 4.4   Loudest of the "young Turks" endorsing Darwin, Thomas Henry "Bulldog" Huxley, in 1874.

how they can go so far, and not embrace the whole Lamarckian doctrine" (DCP 1862). Lyell still found that "unorthodox." But, surprisingly, on 2 May 1856, he nevertheless suggested that Darwin publish his Lamarckian views: "I long to see your application of any modification of the Lamarckian species-making modification system" (DCP 1862). On Lyell's encouragement, Darwin shifted from writing about marine invertebrates to putting down on paper the whole big species origin theory. But his shifting proved, well, glacial. And Wallace got there first.

Ultimately, it was the network of distinguished gentlemen naturalists, including Lyell and Hooker in the United Kingdom and Asa Gray in the United States (Figure 4.5), who would ensure Darwin's dithering wouldn't matter for his scientific standing – his *priority* – as a fellow gentleman naturalist.

Figure 4.5　Harvard University's botanist, Asa Gray, Darwin's devout Presbyterian ally, in 1864.

In-Betweeners

Charles and Emma left London all the way back in 1842, when Emma found a suitable home for them in Downe, a tiny village southeast of London in Kent – not too far, but certainly not too near (and spelled "Down" when they first arrived, which is why they insisted on calling their house, confusingly, "Down House" in Downe). The house was white and large, though not palatial, secluded, but not secluded enough. They had to lower the road out front to prevent casual peepers over their front wall. Behind the house spread an enormous garden and, behind that, perhaps his favorite part: a path through a small wood nicknamed the "sandwalk" where Darwin took daily walks (Figure 4.6).

Figure 4.6 The "sandwalk" – Darwin's thinking space behind the family house at Downe. Little has changed, other than the girth of the trees, from its nineteenth-century state.

Like the barnacles he studied so intently for eight years in his home study there, Darwin cemented himself down, practically immobile ever after, except for excursions to health spas, especially those proclaiming the water-cure. They birthed eight children in that house, plus the two (William and Annie) they brought to Downe with them. They built onto Down House repeatedly over the decades as their brood swelled and they added nannies and other servants (Figure 4.7). Darwin proved an unusually engaged father, it is true, but he and Emma had plenty of help, even by the standards of the day. He invested money, like his father, and did well enough. He learned billiards. He spent nearly all day every day corresponding, reading, playing backgammon, listening to Emma's excellent music, taking pinches of snuff, and having people read to him as he reclined on the sofa. And he wrote numerous geological works; he remained connected to the Geological Society of London after serving as its secretary before moving to Downe.

Twenty years elapsed without any major reason to publish the species book. He put pencil to paper in 1842. Then he wrote out a much longer, neater draft

Figure 4.7 The back of Down House. Charles and Emma Darwin's bedroom was at the far upper bay window. The busy dining room stood two floors below.

in 1844 with instructions to send it to J. D. Hooker if Darwin died. Perhaps after the enormously popular, anonymously authored *Vestiges of the Natural History of Creation* appeared the same year, urging a much more comprehensive evolutionary story than his own, that urgency he may have had to write it back in the late 1830s drained away. He updated the *Journal of Researches* in 1845 without much of a peep that he was thinking about Lamarck and his own grandfather's half-century-old transmutationism, even if Lyell sniffed it out. The large, lavishly illustrated barnacle books didn't mention anything unorthodox. Neither Wallace's 1855 discussion of natural laws leading to speciation nor that illuminating weekend with Huxley and Wollaston in April 1856 quickened his pace. Nor even Lyell's letter to *get on with it*.

We have to imagine that, as he stared at Wallace's letter in mid-June 1858 with its enclosed theory of natural selection, he pondered over his dithering, his delays, his paralyzing insistence that he have rock-solid evidence for every point before publishing anything at all. He spent nearly his entire adult life preparing to shoot his shot. And just before he got that chance, a younger man stepped up and took it.

Scooped. For a writer, especially one clinging to a single monumental idea, few feelings could be worse than being scooped.

But Darwin's years weaving the web of gentlemen scholars were about to pay off. After Darwin contacted Lyell, Lyell contacted Hooker and others. The Linnean Society of London, the beating heart of gentlemen natural historians, agreed to hear the paper of Wallace and several short pieces establishing Darwin as the originator of the theory of descent with modification by Natural Selection. Alfred Russel Wallace was merely a collector and was still in the South Pacific. And Wallace was certainly not a country squire naturalist on his English estate where he could breed his own pigeons, grow his own flowers, and stroll along his own property deep in unbroken thought. Wallace received the honor of having his paper read in front of such august botanists and zoologists at the Linnean in 1858. And he was grateful. Sincerely. He said so in print, and he meant it.

After a year of writing like a man with something to lose, Darwin produced an "abstract" of what he started back in 1837 (i.e., "Notebook B") or 1842 (i.e., the pencil sketch) or 1844 (i.e., the much longer draft) or 1856 (i.e., when Lyell urged him to publish), depending on one's preferred anchor point. After much handwringing and mind-changing, he titled it *On the Origin of Species by Means of Natural Selection, or the Preservation of Favoured Races in the Struggle for Life*, which was published in November 1859. And any notion of "Wallaceism" began to slip away. Even Wallace was content to say so.

Whether or not that course of events proved fair to Wallace, scholars indicate that Wallace's proposed theory of selection differed from Darwin's in crucial ways. When added together, these differences spelled a "Wallaceism" distinct from Darwinism. For many years, scientists and others interested in establishing Darwin's priorities insisted that Wallace's theory covered only groups, while Darwin emphasized individuals. But when read carefully, historians identify nearly as much emphasis on individuals by Wallace as by Darwin and more of an emphasis on groups by Darwin than we used to believe.

Instead, we should see their scientific differences as hinging on several points of *emphasis* and *implication*, which I've listed in ascending importance.

(1) Darwin used domestic breeding as a close analog of what happens in the wild. Wallace disagreed. Domesticated animals didn't need to use the full panoply of wild instincts and so forth. For them, the struggle for existence was muted, the senses of domestic animals dulled.

(2) Darwin stressed the importance of exceedingly slow change; Wallace assumed deep time but didn't address the tempo of evolution. Later, he seemed fine with evolutionary "jumps" or "sports." Darwin's biological gradualism matches up with Lyell's geology, of course. Additionally, it helped Darwin defend his theory against the charge Lyell leveled against Lamarck – given how slow and incomplete is evolutionary change, one couldn't expect to see examples of it over one or even many human lifetimes. Even the seemingly deep look into the past provided by Egyptian mummified animals didn't help – what's five thousand years if the true evolutionary timescale is millions of years?

(3) Darwin explicitly referenced both animals and plants in his natural selection writings; he leaned on Hooker quite heavily for these. Wallace collected invertebrates at a far higher rate than Darwin ever

did and was also a prolific hunter of vertebrates. Plants seemed to matter far less to Wallace, and, given that he didn't have the director of Kew's unparalleled botanical collections at the ready, it shows in his writings.

(4) Darwin's version of natural selection stressed competition between individuals *within* a species for resources. Wallace referenced that but highlighted the struggle for existence as against the environment itself, against predators, *between* species, and against other unrelated organisms that might limit resources.

(5) Perhaps the greatest difference between their versions of evolution is the degree to which each was willing to countenance change that was not immediately tied to survival in the struggle for existence. Wallace seemed to be all-in on adaptation – the *Allmacht* (supreme principle or power) of natural selection is what the Germans called it. "Ultra-Darwinism," as Anglo-American critics briefly dubbed it in the late nineteenth century. As Georges Cuvier used to emphasize from Paris, organisms had to be exquisitely adapted in all their features just to avoid extinction. In Wallace's view, as elegantly described in his 1858 paper, organisms could only change within certain limits. No "unbalanced deficiency" could occur – the scythe of selection operated as a grand regulating mechanism, which he compared to the spinning governor of a steam engine. As it spins faster and faster, the governor dampens the introduction of new steam to keep the engine running at a constant, not a run-away, pace. Only deviations from type that remained adaptive would persist. Darwin's mature theory, though not published until the 1870s, left room for evolutionary change that didn't bear specifically on *survival* alone. I'll come back to that important difference in the next chapter.

Together, these differences proved significant enough to state that Wallaceism and Darwinism were two different conceptions. And then there would be no Wallaceism at all.

Always the Bridesmaid

Other reasons more powerful than intellectual ones also explain why Wallace could never match Darwin's scientific laurels. Class and educational distinctions threw up an insuperable barrier to Wallace's inclusion in the inner circle of Victorian evolutionists. Not surprisingly, given his own often precarious financial status, Wallace advocated more frequently and more openly over time for policies that benefitted common working people like himself. Through the 1860s, he wrote and edited for money, even editing for Charles Lyell and Darwin himself, as if Wallace was a personal secretary.

In print, he disavowed the economic liberalism upheld by some of Darwin's own family members, including Charles and Emma's son Major Leonard Darwin, who would become a low-tax/antiregulation Liberal MP in the 1890s. Wallace, by contrast, converted to Henry George's "single tax" political-economic policies that would have shifted the burden for funding the state onto landholders instead of laborers. The year before Darwin's death, he had gone even farther than mainstream Georgists, advocating for socialized ownership of most land as the first president of the Land Nationalisation Society. Had such policies caught on in government, they might have affected the Darwins severely.

Wallace also stood out in his stance against compulsory vaccination, given the poor quality control of vaccine development in nineteenth-century Britain. Plus, he fulminated against eugenics once Darwin's cousin, Sir Francis Galton, began to agitate for it in 1883. Clearly, just having the imprimatur of the British government or scientific community was not enough to move Wallace. This defiance led to him being pushed further toward the outside of Darwin-promoting circles.

Perhaps more than all of these other factors, Wallace committed a cardinal sin against the Victorian scientific establishment accreting around evolutionary theory that few men recovered from: Wallace endorsed, or at least didn't denounce, Spiritualism.

It started innocently enough. Ironically, Wallace so firmly regarded natural selection as the *Allmacht* (supreme principle) of evolution, a principle that had no equal in biology, that humanity couldn't fit in it. What could be the reason for the enormous mental gap, wondered Wallace, between even the most advanced apes and the humans who lived closest to a state of nature, the hunting-and-gathering Dyak on Mt. Serembu, for instance? Why would any increase in *thinking* power evolve in the first place? And, even if greater mental capacity was slightly adaptive, human thought so often seemed to be adaptive in the wrong direction, *away* from greater survival when compared to the orangutans, *mias*, Wallace encountered in Southeast Asia. Like Lyell had pointed out to Darwin in the past, human intellect – at the sacrifice of stronger muscles, bigger teeth, sharper claws, greater camouflage, an easier birthing process, less proclivity toward depression, addiction, and suicide, and so on – seemed to come out of nowhere, biologically speaking.

In the 1860s, while writing up his *Malay Archipelago* book, and while editing Darwin's and Lyell's work, Wallace hit on an uncomfortable realization. Darwin was writing about humans, which would eventually become the first part of *The Descent of Man* in 1871. But Wallace strongly differed with Darwin regarding human origins. Just as he had a decade earlier, he jumped out in front of the slower, more thorough Darwin, appending his own thoughts on the descent of man to the tail of Wallace's anonymous review of three books by Lyell now in new editions, *Principles of Geology*, *Elements of Geology*, and *Antiquity of Man*, a review that appeared in the widely read *Quarterly Review* in April 1869. Natural selection governed everything in the natural world, Wallace insisted, because everything must be exquisitely adapted for survival in unforgiving, crowded environments. Everything except for humanity, that is. And even while Darwin cranked out his first big book on the subject, Wallace jabbed a defiant flag into the ground in an 1870 book of his own that, once again, beat Darwin to the punch. *Contributions to the Theory of Natural Selection* (1870) insisted that natural selection found its limits in the human head.

> [T]he brain of the lowest savages, and, as far as we yet know, of the pre-historic races, is little inferior in size to that of the highest types of man, and immensely superior to that of the higher animals; while it is universally admitted that quantity of brain is ... the most essential, of the elements which determine mental power. Yet the mental requirements of savages, and the faculties actually exercised by them, are very little above those of animals. The higher feelings of pure morality and refined emotion, and the power of abstract reasoning [e.g., geometry] and ideal conception [e.g., the concept of infinity], are useless to them, are rarely if ever manifested, and have no important relations to their habits, wants, desires, or well-being. They possess a mental organ beyond their needs.

> Natural Selection could only have endowed savage man with a brain a little superior to that of an ape, whereas he actually possesses one very little inferior to that of a philosopher (355–56).

A brain just a bit larger than that of a gorilla could have given humanity everything it needed. What we have instead, Wallace mused, is an imagination machine. Something that delves into secrets at the boundaries of the observable Universe, whether immense or tiny, in the distant future or past, or even things that could never exist in nature (sphinxes, philosopher kings, exclusively charitable and uplifting social media, time-traveling British police boxes, the answer to life, the Universe, and everything, etc.). This apparent gap between human cognition and that of other primates, reasoned Wallace, meant not only that natural selection no longer shaped human minds, but that natural selection, in as much as it requires faithful adaptation to the surrounding environment, could not have originated the human mind in the first place:

> So, those faculties which enable us to transcend time and space, and to realize the wonderful conceptions of mathematics and philosophy, or which give us an intense yearning for abstract truth, (all of which were occasionally manifested at such an early period of human history as to be far in advance of any of the few practical applications which have since grown out of them), are evidently essential to the perfect development of man as a spiritual being, but are utterly inconceivable as having been produced through the action of a law which looks only, and can look only, to the immediate material welfare of the individual or the race (358–59).

He had given up traditional religion long ago. As Wallace himself admitted, deferring to spiritual forces rather than ordinary material ones could only be seen as anathema by many scientists. Yet he thought it completely consistent with evolution by common descent with modification by natural selection; more consistent, in fact, than a materialistic theory of matter that posited only atoms and void. If we adopt a kind of panpsychism – like that proposed by Baruch Spinoza, Gottfried von Leibniz, Margaret Cavendish, Arthur Schopenhauer, C. S. Peirce, and German Darwinist and contemporary of Wallace, Ernst Haeckel, among others – Wallace believed science could more faithfully explain both the connection between substance and force (i.e., there exists only force), and the connection between mind and matter (i.e., mind is a fundamental property of that force).

By itself, this philosophical stance may not have put Wallace at odds with the rest of the biology community gathering around Darwinism. He did argue with T. H. Huxley regarding the importance of cellular protoplasm, which didn't help his cause overall. But that alone wouldn't have been enough to ostracize Wallace. Even placing a tall fence around human intellect, wherein processes that operate on other non-human animals could not enter, likewise proved consistent with other prominent English scientists. "Spiritualism," however, began to mean much more than this for Wallace.

Many years earlier, before leaving for South America in 1848, Wallace became entranced by the practices of mesmerism and hypnosis after hearing that surgeons completed surgeries on patients in a state of hypnosis, and their patients awoke feeling no pain. Wallace wanted to investigate the veracity of these reports. So, he learned hypnosis, even claiming to hypnotize some of his own teenage students at the Collegiate School in Leicester. Upon his return to England from the Southeast Pacific in the 1860s, Wallace returned to these investigations.

Then, at the behest of his sister, Fanny, Wallace also dug into the growing body of literature surrounding communication with the dead. He began attending séances and speaking with mediums. No stories should be rejected out of hand, Wallace believed; everything required some sort of an empirical test. Crystals and magnetic healing, psychic phenomena like second sight and automatic writing, the Roman Catholic phenomena of the stigmata – he assured skeptics that he understood that many people revealed these claims as delusional or fraudulent. But they required actual empirical tests to demonstrate the falsehood. Wallace wasn't about to adjudicate for or against without careful observation.

And then he saw his dead mother. Spirit photography. It was Wallace's breaking point. When he visited spirit photographer Frederick A. Hudson in 1874, he came away after a long photography session (it could be an agonizingly slow process in those days) with three plates showing himself seated with other phantasmic images nearby. The third plate contained that of an old

Figure 4.8   Frederick Hudson's photography trick that ruined Wallace's reputation.

spectral woman in a veil floating near to the camera, in front of Wallace himself (Figure 4.8). When he showed it to one of his sisters, she insisted the phantom was an image of their late mother. Perhaps he held out at first. But soon Wallace came to believe it as well. His long-suffering attempts to remain detached and objective faltered.

He wasn't gullible, exactly: he knew that many of these other phenomena could be shown to be fakes. But not a photograph. Especially since this image wasn't one the family actually possessed of his mother. Wallace admitted in 1875:

> I see no escape from the conclusion that some spiritual being, acquainted with my mother's various aspects during life, produced these recognizable impressions on the plate.

> That she herself still lives and produced these figures may not be proved; but it is a more simple and natural explanation to think that she did so, than to suppose that we are surrounded by beings who carry out an elaborate series of impostures for no other apparent purpose than to

dupe us into a belief in a continued existence after death (Defense of modern Spiritualism, 191n).

The new technology dazzled him. His earnest trust in plain people proved too firm. Claims of forgery piled up, even against photographer Hudson himself, but Wallace persisted in belief. Sure, *some* of these individuals displaying clairvoyance or telekinesis or reporting hauntings or telling fortunes or communicating with the dead during séances could be exposed as con artists. But that didn't mean the entire set of psychic phenomena should be thrown out as being "unscientific." Why would people construct such an elaborate lie? Also, how could any of us know something was unscientific if we hadn't done any science on it? And how do you measure the unmeasurable, see the unseen, reveal what is occult? You consult the experts. But who were the experts on the spirit world?

The problem was that, on the one hand, Wallace distrusted the moneyed, gentlemanly authorities so often speaking in the name of science. It's why he looked askance at Victorian medicine. Physicians and surgeons might talk a good game, and they might receive a great deal of social deference, but their methods proved unreliable, based on nothing more than conjecture even if held with the force of dogma. He was a professional biologist, an evolutionary scientist. Yet medical doctors disbelieved in evolution, scoffed at the latest biological theories. Surgeons were mere technicians, mechanics of wet machines. Why trust them? When it came to smallpox vaccinations, for instance, he didn't. Evolution suggested to him that even microbes had an adaptive role to play in the ecosystem. It would be at our grave peril to disturb that balance, to slice through nature's evolved network of predator, parasite, and prey. And look! People die of badly made, badly administered vaccines all the time, thought Wallace. Even physicians fabricate and bluster. He wanted hard evidence, not just naïve trust in authorities who use Latinate medical words.

On the other hand, Wallace craved the support of other experts. He continued to appeal to Darwin, for instance, year after year, even when they sharply disagreed. Often when scientists and philosophers did attempt to disprove spirit mediums, hauntings, mass visions, and the like, they became convinced that it was *not* faked: those such as physicist William Crookes and philosopher

G. H. Lewes, both of whom went in skeptics and came out believers. Wallace clung especially to their authoritative accounts. That plus his own experience with mediums, spirit photographers, mesmerists, and the like made Wallace still more convinced that Spiritualism contained some grain of truth. Everything is held together by a Force. That Force is also a cosmic mind in which we all are participants, momentary substantiations. It makes sense that we will continue in some form, as ripples in the cosmic ether, for long after our hearts and lungs cease moving.

RIP Wallaceism

Publishing these views – that was his final mistake. "A defense of modern Spiritualism," a too-credulous account he placed in the *Fortnightly Review* in May 1874, started a public fight that would last years. He reprinted it a year later in *On Miracles and Modern Spiritualism*, alongside three essays previously only heard by sympathetic audiences. Other evolutionists dropped their jaws in disbelief. Darwin's circle, including Huxley and Hooker, wanted to cut Wallace off entirely. When 55-year-old Wallace and his family were on the point of financial destitution in the late 1870s, a family friend wrote to Darwin, Hooker, and others. Couldn't one of them help him find a job so he could have some income other than his writings? Many politely ignored the request. Hooker got angry – Wallace had made his bed with the Spiritualists, now let them come to his aid. Only the patience of Darwin changed their minds and generously pulled the Wallaces away from financial ruin. But the damage spread. Racist polygenists arguing against the notion that all human races were biologically of the same species saw Wallace's Spiritualism as evidence that evolutionists were madmen, pseudoscientists blown about by soft-minded fads. Wallace became an object lesson.

His Spiritualist publications threatened the reception of his *Geographical Distribution of Animals*, published in 1876, a book that revolutionized biology and formalized the significance of his biogeographical discoveries in South America and Indonesia. That work earned Wallace true scientific laurels for the rest of his career. We still reference this work today. But the damage was done. A younger Wallace might have seen evidence like that seen by Darwin and reasoned that a Malthusian principle of natural selection must be at work.

An older Wallace might even have gone beyond Darwin defending it. But there would be no "Wallaceism."

And, when a new generation of evolutionary scholars planned commemorations of the publication of *On the Origin of Species* and Darwin's birth (because, conveniently, those events occurred precisely 50 years apart) in 1909, only Darwin's German propagandist, Ernst Haeckel, even mentioned Wallace. It was in a talk that never made it to English celebrants, *Das Weltbild von Darwin und Lamarck* ("The world-construction/plan/view of Darwin and Lamarck"). Not Darwin–Wallaceism. Sadly, Wallace was still alive when those celebrations occurred in 1909. In fact, he had won the first "Darwin–Wallace medal" given by the Linnean Society in 1908, on the 50th anniversary of his paper that the spirits (in)auspiciously delivered from malarial Ternate in the South Pacific to Darwin's quiet study in Downe. But if you attended the 1909 Darwin celebration at Cambridge University headlined by William Bateson, one of a new breed calling themselves "geneticists," you would have heard next to nothing about Wallace.

A *New York Times* interview in 1911 continued the tarring and feathering. "We are Guarded by Spirits Declares Dr. A. R. Wallace," the anonymously authored headline chuckled, despite the fact that half of the article dwelled on Wallace's impressive scientific achievements, and he said no such thing about guardian spirits. By the centenary of the *Origin* in 1959, Wallace was all but forgotten.

In one of those strange paradoxes of history, it was the third son of William Bateson who began to revive interest in Wallace. (Paradoxical because William Bateson himself was a grumpy geneticist who sparred with Wallace and who discounted the importance of natural selection and certainly Wallace's *Allmacht* version of adaptation.)

Gregory Bateson was an anthropologist interested in systems biology and cybernetics who also dabbled with psychedelic psychology and marine animal communication in the 1960s–70s counterculture. And in Wallace's 1858 description of natural selection sent to Darwin, Gregory Bateson deduced something profound. As Bateson recounted it, Wallace presented to Darwin the powerful image that natural selection is "'just like a steam engine with a governor.'" From Bateson's perspective 120 years later, Wallace had enunciated biology's first "cybernetic model," a negative or stabilizing feed-back system containing the explosive growth of organism populations until some variation slipped by the governor, taking the whole species in a new direction. Such notions appeared again in modern computer processes, noted Bateson.

The problem was, as Gregory Bateson's ex-wife, the much more famous anthropologist Margaret Mead, reminded him, Wallace didn't necessarily realize what he'd said. Neither did Darwin. That's why there was no "Wallaceism." Not even Wallace recognized the deep ecological significance of population feedback loops. Looking back on it, however – and looking past Spiritualism and the factors that drove Wallace out to the fringes of the Victorian evolutionary biology community – Bateson insisted, "he'd really said probably the most powerful thing that'd been said in the 19th Century."

"Only nobody knew it," added Mead.

5 "[T]his view of life, with its several powers"

Darwin claimed that *On the Origin of Species, or the Preservation of Favoured Races in the Struggle for Life* was only an "abstract" of that much longer book he had begun to write in 1856, after his irreverent meeting with J. D. Hooker, T. H. Huxley, and T. V. Wollaston, and Lyell's exasperated encouragement in May. But he never completed that larger book. Instead, he worked on plants and pigeons and collected information through surveys from other naturalists and professional specimen hunters like Alfred Russel Wallace for the better part of a decade.

Three years after *Origin*, he published *On the Various Contrivances by which British and Foreign Orchids are Fertilised by Insects*, the result of surveys and some work in his own garden greenhouse. Three years after that, he published a long article, "The Movements and Habits of Climbing Plants," which he published in book form a decade later. And he updated the *Origin* repeatedly through the 1860s. So, he wasn't idle. Yet, it took him until 1868 to finally compose his evolutionary vision with a trilogy written in quick succession: *The Variation of Animals and Plants Under Domestication*, published in 1868, *The Descent of Man; and Selection in Relation to Sex*, published in 1871, and *The Expression of the Emotions in Man and Animals*, published in 1872. It's hard to know what his "big book" on the theory of evolution by natural selection would have looked like had

For the structure of this *Origin* exposition, I'm indebted to lectures given by the University of Notre Dame's Darwin expert, Dr. Phillip R. Sloan, and many years' worth of my students' excellent questions and insights.

he really written it following the *Origin*. But the trilogy helps us see where he saw the holes in his ideas, and how he believed he could shore them up.

Origin of Species does feel like an "abstract," albeit a really long one. When you read it, you encounter hardly any scholarly footnotes, no long tables of observational data (from the *Beagle* voyages, say), and only a single illustration of note – an important one, for sure, but only one. No doubt the speed with which Darwin had to publish something (Wallace symbolized a threat to his scientific prominence and, though he hated to admit it, to Darwin's vanity) contributed to the overall "undocumented" feeling of the text. Perhaps, given that haste, we can forgive some of the attacks by Darwin's first critics. They charged that Darwin merely threw out conjectures "higgledy-piggledy" without sufficient evidence (DCP 2575). Ironically, this was the very charge he heard Lyell make about Lamarck back in his London days in the 1830s. Those critiques extended to grandfather Erasmus Darwin even more. In other words, those were the accusations Charles Darwin most wanted to avoid.

So, what Darwin included in the *Origin of Species* can only be described as a synopsis. And though it concludes with a clear appeal to "several powers" that undergird his process for evolution, it doesn't encompass his complete theory. Unfortunately, that fact leads to our third major misconception, since few people venture into Darwin's works beyond the *Origin* and aren't picky about which edition of the *Origin* they read. Common descent through natural selection was definitely a core feature of Darwin's theory. But not the whole.

In order to understand the rest of it – the goal of this chapter – we need to summarize *On the Origin of Species*, which went through six editions: 1859; 1860; 1861; 1866 (includes "Historical Sketch" of precursors); 1869 (adds "survival of the fittest"); and 1871 (only the sixth uses "evolution"). Plus, we need to incorporate Darwin's theory through *Variation in Animals and Plants* (1868), *Descent of Man* (1871), and *Expression of the Emotions* (1872).

In order to unpack how extensive was Darwin's vision, how different than Wallace's, for instance, let me first dissect the *Origin* itself to see the skeleton of his Big Idea.

On the Origin of Species by Means of Natural Selection, or the Preservation of Favoured Races in the Struggle for Life (1859)

Chapter 1: Variation Under Domestication

The title of the chapter is the key to potentially one of the most important opening moves in the history of science. Darwin creates an on-ramp into his controversial argument through an inoffensive analogy: the breeding of domestic animals including dogs, sheep, cattle, and pigeons. Alfred Russel Wallace saw domestic breeding as an impediment to understanding evolution by natural selection, since humans shield livestock from the full brunt of the environment's demands. But Darwin thought domestication made for a wonderful analogy because of the obvious biological flexibility of domesticated plants and animals. Breeders deliberately choose or select individual traits in one or more organisms leading to more pronounced expressions of those traits. To get more cows' milk, for instance, we would breed those cows that already produced the most, then select from among their offspring those that produce the most, and so on for multiple generations. In the *Origin*, Darwin hopes that the familiarity of domestication means that we, like him, will identify a great deal of variation in these traits in whatever organism we're examining (e.g., snout length in dogs, wool density in llamas, seed density in grapes, needle length of a pine, rose petal color, etc.). His exploration of what happens to, with, and through this variation is a major key to the whole of the *Origin*.

Initially in the chapter, Darwin says that he is interested in the *causes* of variation, and he chalks it up to some sort of a drive, some universal principle at work that makes organisms vary both between parent and child and between siblings from the same parents. Children look different than their parents, no matter if we are talking about mice or men. So, note that we already have a hint that Darwin is going to need a process in addition to natural selection. But rather than dwelling on the causes of that variation, that universal principle of differentiation, Darwin waves his hands and says we should just take it for granted that it exists. The best he can say is that reproductive elements were affected before reproduction. It would take him a decade to come back to the topic in *Variation in Animals and Plants Under Domestication*.

His insight is not new, of course – Aristotle wrote about variability in organisms many, many centuries ago. Darwin asks us to think again about the domestic breeder trying to get more milk. Now speculate back in time: Are all traits just additions of earlier ones? Did large and hairy St. Bernard dogs come from combinations of just a bit larger, somewhat hairier wild dogs? Did tiny toy poodles come about from combinations of much smaller than average wild dogs? Does this mean that all things we call "dogs" were originally from one set of common ancestors, then? Did the variation that produced toy poodles and St. Bernard's all exist in some original animal?

In asking this question, Darwin has snuck in a much more challenging question, the one he and Huxley and Hooker and Wollaston chuckled about back in 1856, namely the whole problem of taxonomic classification at the species level. *What is a species?* Are there "natural" species, meaning species that nature has set apart as distinct regardless of how taxonomists classify them? Are all domestic dogs part of a single "aboriginally distinct species" (that's Darwin's term)? A whole century earlier, French biologists claimed that there was only a single criterion that could decide this question: Can these organisms produce fertile offspring? Darwin will dodge this question here, too, just noting that the fertile offspring species concept is not an all-or-nothing proposition. Given that, he concludes that the fancy pigeons that he breeds himself must have come from an original pair of rock doves. But dogs couldn't have come from just one progenitor. (I mean, really, how could a pug and a wolfhound share a recent common ancestor?)

Next, Darwin takes on a potential suspicion voiced by Lyell and others: Is there some limit to variation? Can human breeders, for instance, bend an organism so far that it becomes another kind of thing altogether? Notice that he is interested in the choice of breeders. He is not confident, after talking with a fair number of professional breeders, that domestic forms have not developed from a varying background of climate or general environmental changes. The variation already existed, and the breeders picked it out – believing that almost any kind of change is at least possible in principle.

Darwin's real reason for pointing this out, however, is to introduce the concept of "unconscious" or implicit selection by breeders. Perhaps selecting for more cow milk also selected for taller cows or cows with more spots. Perhaps in selecting smarter and more muscular German Shepherds with sloping backs, we also unconsciously "selected" for canine hip dysplasia. Darwin is convinced that this kind of unconscious selection has, overall, resulted in a continual and general improvement of domesticated forms without anyone intentionally altering them for more specific traits.

At this stage, we must agree with Darwin that unconscious selection by breeders from time immemorial works for the general improvement of domestic animals if we want to go further with him in his argument. If we don't think there's unconscious selection (but of course there is), then we can't buy any argument about natural selection.

And that brings out another feature of this work. Darwin does not structure the argument in the *Origin of Species* like a prototypical physical scientist, like, say, the gold standard: Isaac Newton's *Principia Mathematica*, published in 1687. *Origin* does not begin with a clear set of axioms and inviolable laws or proofs of statements. All he is doing is setting out relatively modest claims about an everyday practice with which many educated Victorians would have been familiar. He will extrapolate most of the rest of his argument from this analogy. In 1790s, the philosopher Immanuel Kant said that there would never be a "Newton of a blade of grass." More recent supporters of Darwin claim that he was biology's Newton. But by the close of the first chapter of the *Origin*, readers could find little evidence for that. There's just too much homespun analogy at work here.

Chapter 2: Variation Under Nature

If domestic breeding served as the fulcrum of that first chapter, this chapter pulls the analogy over into the wild world. Immediately, Darwin launches into the argument over what constitutes a "species." It's just really hard to say what a species is, he says. That's strange for someone who worked on a classification project for years and years; except if you reflect that Darwin was working on barnacles, invertebrates, like Lamarck specialized in – but not Cuvier or Richard Owen or the better-known comparative anatomists who all worked with vertebrates.

Darwin then begins deploying similar arguments for "variety." However, one thing is definitely different about his use of variety versus species.

Varieties imply a pre-existing "community of descent," he says. We can see their relatedness with our own eyes. And then he moves toward undermining the differences between varieties and species. If there is no difference (except in the mind of the taxonomist) between a variety and a species, and if varieties show common descent, then species, too, should show signs of common descent.

But how will he make this argument convincing? I mean, he can say "common descent" all he wants, but how will Darwin *show* this?

He starts with his own experience. About halfway through the chapter, Darwin lets us see some of his own frustrations as a naturalist: "When a young naturalist commences the study of a group of organisms quite unknown to him, he is at first much perplexed in determining what differences to consider as specific and what as varietal...." You can imagine Darwin's experience classifying barnacles in the 1840s–50s. Squinting through his magnifying lens, all he saw was a huge number of traits. Which ones were common to all these organisms? Which ones were just the everyday kinds of differences? Which traits are inclusive of some subset of organisms but not others? This might not be tough if you're breeding livestock or dogs. But imagine barnacles or earthworms – the process is frustratingly difficult. Darwin takes this difficulty seriously instead of assuming it's because the naturalist is unskilled. An ordinary professional taxonomist might see some small individual differences and shrug them off, he admits. That's because the taxonomist is looking for similarities in order to group individuals into larger taxonomic units. By contrast, Darwin celebrates individual variation as "of the highest importance." Variations at the *individual* level become, Darwin claims, a variety – variations that persist among a small number of individuals in a location. The group variety, if the organisms with those variations thrive and enter more areas, could become a sub-species. Eventually, it's possible that organisms possessing these variations become full-blown species. Darwin connects all these dots together: "A well-marked variety may therefore be called an incipient species," he says, as if this is no big deal.

Actually, this is a shocking statement. *Incipient* literally means "in the process of becoming." Darwin reveals that he believes that varieties are embryonic species and that there is *no unambiguous way* to tell whether a variety (a category that we cannot truly define either, as Darwin just said) is *not* going to become a real species that all "good" taxonomists will agree upon. (Yes, that's a double-negative.) He emphasizes that this whole classification business is a whole lot murkier than it appears to non-specialists, that it's tough to make confident claims about any of it. So, if we're honest, we should admit that there's lots of room for sliding around between kinds of organisms – it's just that hard to tell kinds apart. More importantly, because he says that varieties imply "communities of descent," and there is a line between varieties and species, Darwin just made taxonomy about historical communities of descent – relationships between past organisms living through past events and their present descendants. Long-dead progenitors reappearing through traits possessed by their great-great-great-etc. grandchildren.

Again, just as we saw in his chapter 1, Darwin leans pretty heavily on analogies. At minimum, he asks us readers to stretch arguments and evidence that pertain in one localized case (e.g., the barnyard) to much broader situations (e.g., wild spaces). Creationists jump on this sliding analogy business today. However, Isaac Newton himself argued that this kind of reasoning was completely legitimate. Though he could not actually observe the effects of gravity on distant stars, Newton generalized his observations of attractive forces on Earth to all bodies whatsoever in the Universe. In Darwin's own day, polymath William Whewell, one of Darwin's philosophically and scientifically astute professors at Cambridge, agreed that one could rationally make these sorts of generalizations from specific cases to general ones through a kind of strict ladder of inferences. That's precisely what Darwin attempts to do in his chapter 2: species within a genus are related to each other just as varieties within a species are related to one another.

Chapter 3: Struggle for Existence

As important as chapters 1 and 2 are, 3 and 4 are really the heart of the *Origin of Species*. Understanding these two chapters, then, is crucial for understanding Darwin's vision and the future of evolutionary theory in the nineteenth century to the present. Right at the very beginning of the chapter, Darwin says that the "struggle for existence" – discussed by Malthus and Harriet Martineau as well as other naturalists – has something to do with the selecting process. It will take the rest of these next two chapters to figure out what that relationship is.

In this chapter, Darwin raises a significant causal question. In the previous chapter, he asserted that "varieties" could be considered "incipient species." Now he will tell us what causal process lies behind such a statement. Note how important this argument is. Everyone recognized that organisms within a species fluctuate. So what? Maybe they also fluctuate back to their "mean" or "type"; maybe they "revert." Darwin will say that, no, this process only works to diverge – to push outwards – never to converge.

But how? Darwin sounds like he might be thinking of Newton's (and before him, Galileo's and Descartes' – and on back to Jean Buridan in the 1300s) concept of inertia: entities in motion stay in motion until acted upon by another force. Malthus's population principle does the same work for Darwin. Organisms, he says, reproduce to fill every available space unless "checked." Of course, he cannot show that this tendency is in fact a principle adhered to by all organisms, plant, animal, and otherwise. But if you accept this as a fundamental principle, like inertia, you have a bedrock concept in Darwin's theory.

Here is one of his "several powers" that make evolution work. Reproduction with increasing variation is the "push" force behind Darwin's organic machine. A push from inside, just like Lyell identified in geology.

Darwin's job after this point is not to explain the push force but to explain why we do not see the world overrun by any particular species. (Granted, some people would say there are too many humans; I would argue there are too many mosquitoes and poison ivy vines.) Like in the case of inertia, we must assume with Darwin that some other "force" – a pull force or a resistance force, something vaguely akin to Lyell's erosion force in geology – is keeping a lid on overpopulation. Our culprit is this "struggle for existence."

Climate, food, and disease are obvious "pulls" on organic overproduction, just like Wallace pointed out. But Darwin wants us to think more about inter-actions between individual organisms. Darwin paints a picture of a web of relationships, taut guidewires, with organisms pulling on other organisms, keeping them from overpopulating. We're now set up for his breakthrough (quite literally) idea.

In the first chapter, he presented us with an analogy about intentional human breeding; in the second chapter he then stated that this same situation could play out without an intentional agent ordering the breeding. Also in chapter 2, Darwin undercuts any suspicion that varieties and sub-species are "real" things and then states that the species designation is no better than that of varieties. Here he gives the first of his forces, a push force – constant positive increase in population and individual variation – and the second force, an equal and opposite resistance force – limitations in resources and the webs of relationships between predators and prey, plants and animals, parasites, and even seemingly unrelated species.

Except it's not *always* equal and opposite. As he's about to reveal, sometimes the "push" of diversity overcomes the "pull" of competition for resources.

Chapter 4: Natural Selection

This is the core. In this chapter, Darwin condenses that tree from the "Zoonomia" notebook (B) and almost everything else he had worked on up to Wallace's letter in 1858 into one compact thesis. As we might expect, this is the chapter that will receive the brunt of the various criticisms over the next few decades. (If you want to see how these debates shaped this chapter, examine the 1859 first edition alongside the 1872 sixth edition.)

As he so often did when trying to be persuasive, Darwin weakly backs into his argument:

> Can it then be thought improbable, seeing that variations useful to man have undoubtedly occurred, that other variations useful in some way to each being in the great and complex battle of life, should occur in the course of many successive generations? If such do occur, can we doubt (remembering that many more individuals are born than can possibly survive) that individuals having any advantage, however slight, over others, would have the best chance of surviving and procreating their kind? (80–1)

Is it *improbable*, emphasizes Darwin, that helpful variations might allow an organism to escape that "pull" force ever so slightly? Of course we would not find this improbable – it already happens in agricultural breeding. If this happens, Darwin continues, then is it *improbable* that those organisms would leave behind more and more of their kind until they completely replaced their ancestral species after thousands of generations? Note that at each point, he's forcing us to make a counterclaim that it is completely *improbable* that this could happen. We don't have any real evidence to doubt him, though. So, we're going to have to keep going with him.

Darwin moves stepwise through his very simple argument here: (1) variation occurs; (2a) humans select variations that benefit us; (2b) nature selects variations that benefit that organism; (3) variations that confer some benefit will allow the possessing organism to escape some of the "pull" resistance to the "push" of population growth; (4) eventually the groups of organisms possessing these beneficial variations replace the ancestral groups that do not. Natural Selection, which he names here (and capitalizes!) is nothing more than "differential preservation" of the organisms that bear beneficial traits (81).

Wait, skeptics might say, *humans* do the selecting of agricultural organisms. We *allow* the "push" of variation to overcome the "pull" of selection. Are you telling me that an unconscious combination of push and pull forces does the same work?

Darwin jumps back into his analogy between agricultural selection and natural selection once again. Exactly! he says to his skeptics: we would never question that humans select on "external and visible characters" (83). Would we disagree with Darwin that something as powerful as *Nature* could see even further than humans? Nature can "act on every internal organ, on every shade of constitutional difference, on the whole machinery of life" (83). Humans are only in it for extra milk or wool or leather or timber or rose petals or whatnot. Nature, by contrast, wants what is best for the whole organism. At least in this first edition of the *Origin* – though later editions will try to explain away this very anthropomorphic language – Darwin will refer to Nature "scrutinizing," "silently and insensibly working," and acting for the "good" of each organism. Very curious language if one is merely speaking of a tendency or force, such as inertia or gravitation. Perhaps this is some of the old natural theology language of Henslow and Sedgwick and, behind them, Paley, shining through. (I'll tackle that below in Chapter 6.)

Readers who approach the *Origin of Species* familiar with how natural selection is commonly understood today might think, "Okay, this is the point of the whole thing; you can stop now, Darwin." But if he stops now, Darwin is in danger of offering merely a tautological statement: organisms persist because they are best suited to persist. This critique is, in fact, one that will be leveled at Darwin's theory again and again. But there's more going on.

Darwin hints back to Cuvier's coordination of parts argument. As I mentioned back in Chapter 2, Cuvier, the opponent of Lamarck and grandfather Erasmus Darwin, argued that each organism had to be considered as a whole, not in terms of isolated systems or traits. The important point for Cuvier was to show how *extinction* worked. Conditions of existence change so much that the harmonious organization of an organism cannot adapt, because it cannot change whole-cloth. Charles Darwin does something interesting with Cuvier's argument. He insists on slight advantageous modifications, perhaps only of a single part. But Cuvier was right: even slight modifications mean a broader reorganization of sections of organisms. Larger teeth mean a more robust jaw plus jaw muscles plus zygomatic arches on which to anchor those muscles. Very slight changes in one organ might mean very slight coordinated changes in a number of connected organs and tissues. But thinking back to those birds he preserved on the Galápagos decades earlier, Darwin perhaps recalled that John Gould found that even the finches with very thick beaks proved to be otherwise quite similar to finches with smaller beaks. They shared much of the same plumage, for instance – presumably a factor less important to survival than beak structure! So, while the struggle for existence is intense, parts of the organism might change in ways that help it to escape the pull force and additionally change other incidental parts (or not). Adaptation might not be equally applied to all parts of an organism all the time. In other words, potentially, there could be "neutral" change in one part tied to adaptive change in another part. Keep this in mind: neutral change – meaning not improving survival but not hindering it, either – will be an important feature of his theory that Darwin will explore in books after the *Origin of Species*.

But there's an even more significant point here. It's what we now call co-evolution, the cornerstone of modern ecology. Imagine a single flower-bearing plant could produce a slightly greater amount of nectar than its neighbors. All other things being equal, insects would be likely to favor that plant. If the flowers were also arranged in such a way that the visiting insect would pick up pollen while feeding, only to deposit the pollen on the pistils of the next plant, the plant secreting the slightly greater amount of nectar would tend to be fertilized more often. Eventually that type of plant, if future generations kept over-producing nectar, would supplant the other forms. And here's the really cool point that Darwin also sees: the insects able to more efficiently exploit changes in that plant's nectar production would also tend to dominate over generations. The two organisms would *co*-evolve.

Between the possibility of neutral change linked to adaptive change and co-evolution of two organisms, we can now see how much larger and more complex Darwin's vision is (versus, for example, Wallace's). It's not enough to say that selection works on an isolated animal or plant to fit a single set of environmental conditions. Rather organisms gradually alter in tandem (note that Darwin doesn't show that these directly alter each other, just that they are being altered at the same time), even in ways that don't seem completely responsive to environmental conditions at first.

The Diagram

We've now covered the key parts of Darwin's theory of natural selection as laid out in the *Origin*.

There are some problems here that Darwin is going to paper over for the moment. Some turn out to be so important that he'll devote the rest of his life trying to address them. In so doing, his theory will grow far larger and more elaborate than what he sketches in the first edition of the *Origin*. We'll return to the problems, his fixes, and why it's a misconception to see merely natural selection as his major contribution to evolutionary theory below. First, we need to examine his chapter 4 diagram (Figure 5.1).

This is the only diagram that appears in the *Origin* in any edition, so it must be important. Admittedly, however, it's not very interesting. At first, Darwin asks

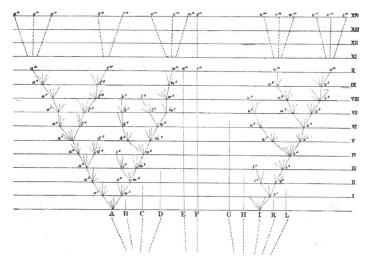

Figure 5.1   The "divergence" diagram in *On the Origin of Species*, unchanged through all editions.

us to think of this diagram as a graphical representation of historical relations at the species level. Individual branches represent varieties in a species. He also asks us to assume that the horizontal lines represent durations of limited time – perhaps 1,000 generations – stacked much like geologic strata. The horizontal axis itself represents the degree of divergence. Looking at the graph as a whole, we can see that, over time, varieties (he labels them "A" to "L") diverge from one another. That's the push force – the principle of divergence; organisms becoming more different over time. And, as he indicates in the graph, ancestral forms almost invariably disappear, leaving in their place new varieties.

So far, Darwin has not said anything terribly controversial. If we think about cattle breeding, we can imagine many breeds of cattle, most of which have diverged over centuries through the action of breeders to produce more milk, meat, smaller horns, and so on. The originals that were smaller in stature and produced less of the products humans want exist only as stuffed animals in museums now. Perhaps the chart is just a graphic representation of that.

But no! Darwin asks us to make a great leap. In his chapter 2, Darwin argued that we cannot decide once and for all the difference between merely a variety and a genuine species. Here, Darwin argues that the scale depicted by the graph is completely relative. If the time coordinate is small, we should see the chart depicting only the minor formation of varieties. "But," he says here, "we have only to suppose the steps in the process of modification to be more numerous or greater in amount, to convert these three forms into well-defined species." Now it makes sense why Darwin would argue that varieties and species are indistinguishable, more or less. The formation of new varieties from a parent variety and that of new species from an ancestral species are the same process. Only the timescale needs to be changed. That is why he can see this single graphic depicting a schema for the *entire* history of life, including living and fossil species, genera, families, and so forth.

In making this logical jump, Darwin is hinting at his interest in articulating a biological law, similar to Newton's laws in physics. If indeed he hoped to be Newtonian, then we should examine how effective Darwin was at following the method proclaimed to be Newton's by Darwin's philosopher of science colleagues (specifically William Whewell, who Darwin quotes in the frontispiece of the *Origin*).

In his famous *Philosophiæ Naturalis Principia Mathematica* (The Mathematical Principles of Natural Philosophy; usually shortened to *Principia*), published in 1687–89, Newton lays out axiomatic laws and corollaries from which he derives a theoretical structure. For example, he derives from his axioms and corollaries the laws governing the motion of an ideal body moving in an ideal circle. Then he transfers this movement to an ellipse or oval. Eureka! This looks like the orbits of planets and coincides pretty well with the laws governing the motions of planets postulated decades earlier by astronomer Johannes Kepler using data meticulously collected by Tycho Brahe's Danish astronomers. In Book 2 of the *Principia*, Newton explains orbits using the old notion of gravity, once of interest only to astrologers seeking to explain how stars, planets, and the moon govern human behavior. Newton extended gravity to cover everything on Earth as it is in the Heavens. In *Principia* Book 3, Newton shows that the observed phenomena fit another recognized mathematical principle: the inverse-square law – as objects get twice as far away, the strength of the force isn't half, it's one-quarter; four times the distance makes the force only one-sixteenth as strong, and so on. When Newton gathers the observed motions of moons orbiting Jupiter, for instance, he sees that they conform to expected results (excusing small errors) based on that inverse-square law. And so does the motion of our moon around Earth.

So, if we wanted to follow Newton's example, the best science, in other words, we would want to first identify an ideal situation, then both the mathematics to describe and the laws that explain that ideal. We would have to make a precise prediction. Then we would look at meticulously collected observations, hopefully a bunch of them, to compare the expected results with what is actually observed.

But this is just what Darwin *doesn't* do. Instead, he takes his argument in the *Origin* in two different directions. First, he challenges his critics (real or potential) to produce a *counternarrative* – if you disagree with him, you'll need come up with some other hypothesis how this works, aside from "God just made it this way." Second, he argues that only the theory of descent with modification can bring widely divergent classes of phenomena under a few simple principles – a concept called "unification" (or, in some cases, "reduction" or "consilience"). By doing this, Darwin suggests, we will have better scientific understanding of processes and entities otherwise unintelligible to us. In other words, he doesn't mathematize, he doesn't make predictions in the abstract and attempt to find real-world cases to test. Instead, he puts out a plausible explanation for why the world appears the way it does and invites challengers to come up with something better. And then, just when a skeptic might find some smaller detail to take issue with in Darwin's theory, he shows a different set of phenomena that his idea *also* explains.

Yet, if he's not really doing what scientists since Newton believed was good science, is his method deficient? Or does biology obey different sets of rules than physics and chemistry?

Problems

Hold on, though. I mentioned above that there are real problems here. In fact, critics immediately found two big holes in Darwin's theory. Note that behind Darwin's entire project sit two critical assumptions: (1) these small variations are heritable, and (2) the variations accumulate in the same direction. But here's the thing: (a) Darwin doesn't have a developed explanation of inheritance, so he must appeal to the unknown laws of inheritance he referenced in chapter 1. And (b) it's not obvious that favorable variations are additive. Perhaps variations fluctuate randomly within certain limits that remain fixed for a given species. That's certainly what Lyell thought. In our flower and insect example, above, it's plausible that a given species of flower could never be more than slightly more productive of nectar than the parental form and never to the degree that the structure of the flower, let alone the insect, would change in a noticeable way.

In fact, here's one of those points where Wallace seems to break with Darwin. In his 1858 communication announcing his own theory of natural selection, Wallace suggests that a change in one part of an organism must somehow be *offset* by a change in another part. That's what Cuvier said decades earlier as well. Organisms have kinds of feedback loops that constrain the amount of change possible, until they somehow overcome these constraints – though Wallace was as unclear about how precisely this could happen as Darwin was.

So, unintentionally, Darwin presents readers with a methodological conundrum: should science proceed by making conjectures only when there's a large build-up of evidence (the technical term here is "induction")? Or does science first make imaginative leaps about the way the world should work and then "interrogate nature" (the technical term is "deduction")? Not only is this an important issue in general, but it is this point more than any other that motivates criticisms of the *Origin*. If Darwin's theory cannot be held to general standards of empirical testability used in, say, physics, is this a problem in *his* argument alone or a necessary consequence of *all* theories dealing with issues that cannot be directly tested by medium-sized humans with only a few-decade-long careers and stuck on planet Earth? To put it another way, how much can scientists say reliably about the deep past, or about the fringes of observable space, or about the exceedingly small, the exceedingly large, or the very temporary?

Darwin tried to ameliorate skeptics by appealing to the same kind of *uniformitarianism* that geologists such as Lyell had used in their theories of the Earth: all things being equal, past forces and phenomena are the same ones working with the same intensity in the present, such that observations done today help us to understand phenomena hundreds, thousands, or even millions of years ago. Just as Lyell had denied sudden changes in the geological realm, Darwin denies them in the biological. And just like we cannot watch mountains grow with just our eyes and our very brief human lifespans, we cannot watch variations become sub-species, become species, and so on. In Darwin's day, it would have been nearly impossible to find any simple verification of his thesis by a test or an observation within a single human lifespan. But for Darwin, this isn't a problem, really. Extrapolation backwards in time from present observations is a legitimate scientific technique, according to Lyell.

Another problem of which Darwin is aware regards the problem of blending. If a new and evolutionarily special trait does occur, why would it persist, rather than being overwhelmed by all the "normal" individuals available for the "special" individual to mate with? Darwin emphasizes processes that would isolate populations of organisms. Perhaps different populations occupy different stations. One species of fish might have a surface-feeding variant and a bottom-feeding variant. Or what about the cases of near-total geographic isolation, such as tortoises on the Galápagos Islands? Isolation might severely limit the numbers of individuals with "favorable" variations. But it might also create conditions in which organisms could persist for long periods of time without introductions of new creatures to change the conditions of existence (think, marsupials in Australia). Natural selection brings about change only if there is a significant enough change in the environment. Wallace hammered home this point, more than any other, and Darwin emphasized it throughout the *Origin*.

In summary, then, Darwin's initial theory unrolled in *On the Origin of Species* advocated common descent by modification given a process mirroring Wallace's, but with similarities to older ideas.

(a)   The rapid increase of organisms in any given environment leads to a struggle for existence – too many mouths to feed, given the resources available.

(b)   Inherited variation, *due to unknown causes* but visible between siblings, with some varieties adapting better to environmental conditions than others.

(c)   Over time, those who adapt the best, given their variations, will continue; those who do not will perish.

(d)   Universal, unceasing gradual change of environmental conditions means that varieties will be joined or replaced by new varieties until the distinctions between members of two related groups will appear large enough to be taxonomically important – in other words, a new variety, sub-species, or species will have come into existence.

These are Darwin's cornerstone principles. Darwin spends the remaining chapters of the *Origin* elaborating on them and demonstrating why these are the driving forces behind descent with modification. It's already a good deal more advanced than Wallace's account. But a decade later, Darwin will push his vision farther, attempting to shore up those problems.

One final point about the *Origin*. Note that the subtitle is *"the Preservation of Favoured Races in the Struggle for Life."* Why did he not choose "variation"? Is "race" just a synonym for "species"? If natural selection works on individual traits and the organisms with those traits, how can we confidently say that a whole "race" is the thing preserved? Does Darwin just mean "individuals"? As these questions simply about the subtitle indicate, a great deal more needs to be unpacked before we will have a true grasp of Darwin's theory beyond "common descent with modification" – now a well-known position. It will take Darwin nearly a whole decade to take that next step.

The Variation of Animals and Plants Under Domestication (1868)

Given our modern prejudices, we likely believe the *Origin* exploded in popularity, given that it represented a major social bombshell and was the first major scientific treatise on evolution. Yet, as you know by now, it was neither of those things. Darwin's book didn't even out-sell other evolution books at first: the anonymous *Vestiges of the Natural History of Creation*, published in 1844, advocating evolution on a cosmic scale, sold far more copies in the short term. For eight years, with his children mostly grown and mostly on to other pursuits, Darwin puttered in his house and large back garden at Downe, concentrating on correspondence, updating *Origin* regularly to address various criticisms.

In the mid-1860s, Darwin pivoted toward addressing the two major problems introduced in (b) above. How did variation occur? How was variation inherited from one generation to the next? This is what he tackled in the two volume *The Variation of Animals and Plants Under Domestication*, published in 1868, an extended dive into the topics covered by the first two chapters of the *Origin*.

Darwin again relied upon the principle of divergence to explain how, in an organism governed by the coordination of parts, change could still occur. In a sense, organisms are bound together by "pushes" and "pulls," positives and negatives, attractants and repellants, just like populations. Every organism experiences an internal push to diversify and – as anatomist Cuvier maintained a generation earlier – an external constraint to remain coordinated with the rest of its parts. (Malthus proposed the population growth opposed by "checks" to keep population growth from overwhelming all, though it was Wallace who most clearly envisioned this feedback system.) At the individual level, the looseness of this connection between the internal push and pull is what provides variation. For the majority of two drawn-out volumes, Darwin provides example after example of how variations and species still show the marks of common descent. Many of these examples come from observations of his own dogs and cats plus surveys of animal breeders across Britain and the Continent. Darwin also explores in minute detail his own work on the hobby of pigeon breeding conducted in his own back garden (Figure 5.2). Repeatedly, he cycles between descriptions of individual variations and anatomical similarities. The goal, it seems, is to hammer *common descent through slight modifications* home.

That's fine as far as it goes. But what upsets this balance such that truly *evolutionary* change is possible?

Surprisingly for modern audiences steeped in neo-Darwinism formulated by biologists in the twentieth century, Darwin goes back to Lamarck and even to

Fɪɢ. 70.—English Barb.

Figure 5.2 Darwin's *Variation of Animals and Plants Under Domestication* featured homely breeds Darwin knew well, like this English Barb pigeon.

his own grandfather's theories expounded in *Zoonomia* and the appendices of *The Temple of Nature*, published over a half-century earlier in 1804. Variation results from the use or disuse of organs, Charles Darwin says. Or environmental conditions directly change organisms. Selection preserves those variations that best fit the new environmental conditions. So, Darwin's Darwinism is really natural selection stacked on top of older ideas about the inheritance of acquired characteristics.

But how, exactly? This criticism stood among the most potent of those leveled against Lamarck's transmutationism, even by Darwin himself. To explain how amphibians could develop lungs when ancestral fish had gills, or how cave fish could lose their eyes over generations in the permanent blackness, Darwin dredged up a French theory pre-dating even Lamarck: "pangenesis."

For 50 pages of *Variation*, Darwin speculates that particles called "gemmules," scattered throughout the body, swell and swell during development until they become the visible parts of an organism. Something akin to pomegranate seeds growing into organs and limbs, I suppose. Of course, those slowly swelling gemmules receive various inputs throughout their existences – bumps and nicks and scratches and scars, mostly; maybe direct, even traumatic, injuries. Some of these gemmules, now altered through the process of being kicked around by life for a while, separate from the others and gather in the sex cells. From there they have a chance to transmit their combined inherited and directly acquired characteristics to the next generation. So, from Darwin's perspective, the biological fact of variation, which was crucial for his concept of natural selection, emerges as little parcels of inheritance, scuffed up a bit, delivered to the next generation (Figure 5.3).

Darwin was right to worry about how this would be received. Pangenesis was an idea too old and out of style by the 1860s to even to call it retro. Pierre-Louis Maupertuis published a thorough study of the concept a whole century earlier. In 1865, three years before publishing *Variation*, Darwin floated it to ardent supporter

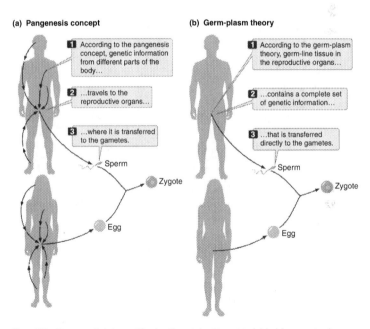

(a) Pangenesis concept

1 According to the pangenesis concept, genetic information from different parts of the body…

2 …travels to the reproductive organs…

3 …where it is transferred to the gametes.

Sperm

Zygote

Egg

(b) Germ-plasm theory

1 According to the germ-plasm theory, germ-line tissue in the reproductive organs…

2 …contains a complete set of genetic information…

3 …that is transferred directly to the gametes.

Sperm

Zygote

Egg

Figure 5.3   Pangenesis (a) was Charles Darwin's attempt to fold older acquired characteristics theories into his natural selection process. The competing germ-plasm theory (b) did not allow for external inputs into a seemingly eternal, discrete "germ," later called "genes."

T. H. Huxley. Huxley read Darwin's pangenesis manuscript carefully and responded to the scion of evolution deferentially, as usual. But Huxley was not impressed. Among other critiques, he gently reminded Darwin that figures like Maupertuis, known to almost all practicing biologists, hadn't managed to convince anyone even back then (save perhaps grandfather Erasmus Darwin). If anything, Huxley thought, matters had only grown more complicated in the nineteenth century with the invention of more powerful microscopes. The problem was, Darwin's gemmules were nowhere to be seen. It would be better for Darwin to avoid the subject entirely or to try again, to come up with something better, suggested Huxley. A little sheepish, Darwin promised to *try* not to publish it. But three years later, here pangenesis sat, splayed out in print for everyone to gawk at.

Huxley remained skeptical even after Darwin's *Variation* hit bookshelves sporting the pangenesis hypothesis. But for the most part, he kept his jaws clamped shut. Other critics were louder. The concern raised upon publication of the *Origin* a decade earlier resurfaced: that despite chapter after chapter of examples of animal and plant breeding, Darwin had relied too much on imaginative speculation rather than hard evidence gathered by rigorous induction when it came to pangenesis, the actual mechanism that made the whole system work. What's more, it was painfully evident that the great naturalist Darwin had barely been able to keep up with the latest microscope-based science. Had he seen a gemmule? Had anyone?

The hypothesis of pangenesis felt threadbare, but it was necessary. Darwin needed some hypothesis for how natural selection was possible. Otherwise, what was the internal "push" force that got "pulled" by selection? What else could produce that "principle of divergence"? Where else could variation come from? So, when it came to this inheritance of acquired characteristics stuff, the cause of variation from which nature carefully selected to improve the species little by little, Darwin had to commit.

The Descent of Man; and Selection in Relation to Sex (1871)

He kept avoiding humans – the subject, that is. He suspected this topic would stir up trouble; though he also confessed, when embarking on the "chapter on man" of *Variation in Animals and Plants* that eventually swelled like gemmules to become *Descent of Man* three years later, that "abuse is as good as praise for selling a Book" (DCP 5384). But Wallace beat Darwin to the punch yet again. This time, it was in 1866 with Wallace's essay delivered to the polygenists of the Anthropological Society of London. He told that relatively hostile audience that human races indeed share common descent with one another (monogenism), at least in body, though not necessarily in mind. In fact, *so many* people had already published on the evolution of humanity by 1870 that 60-year-old Darwin looked conspicuous by his absence.

The problem was monogenism itself. Ironically, the evolutionary notion that humans shared common descent with non-human primates as well as each other also seemed to track with old scriptural accounts of an original Adam and Eve giving rise to humanity. Polygenism – all human races have different origin stories – seemed more scientific by the 1870s. It certainly comported with the oldest evidence then available. Archaeologists found remarkable consistency between the racial characteristics of the ancient civilizations of North Africa and the Near East and today's occupants of the same ground. Humans hadn't varied at all in many millennia.

In his 1866 paper, Wallace tried to bridge the gap between the monogenists and the polygenists by conceding that, while human *bodies* were all related, human intelligence, emotions, and moral systems were not really. Darwin, by contrast, couldn't see the reason to disaggregate human minds from bodies. There were far more commonalities between humans and non-humans mentally than people like Wallace wanted to permit. It stood to reason, then, that all human races were related, body and mind.

But how to explain the differences between human races? Indeed, just as the subtitle of *Origin* specified, evolution happened by the preservation of favored *races*. It would be easy enough to pin all variation to use and disuse, pangenesis, and whatnot, like he had done in his previous work. The more he gathered survey data and actual samples of birds and mammals from around the world, however, the more that he saw a second, overlapping challenge. Sexual dimorphism: the reason why men and women look so anatomically different.

In some non-humans, sexual dimorphism barely exists. But in birds – his pigeons, for example – coloration, song, even feather size and number varied wildly between the male and females of certain species. Look at the extreme differences of peacocks and peahens, for instance. He saw something similar among some species of barnacles that he studied 15 years earlier. While females could be full, active individuals, males sometimes served as little more than lethargic bags of sperm (male barnacles, that is). Just like grandfather Erasmus said two generations earlier, Charles Darwin saw it in the massive horns of large male mammals, the songs of male birds, the flight patterns of male fireflies, the size differences between male and female spiders, even in the courting displays by marine creatures (Figure 5.4). The funny thing is, Darwin mused, some of these features are *anti*-adaptive. Bright coloration, ostentatious behaviors, big floppy feathers, loud calls – all of these things make you more likely to be prey.

Was there one silver bullet that could account for all these features? Could evolution explain the panoply of human races, peacock feathers, silvering on the backs of male gorillas, the squawk of male macaws, bighorn sheep smacking heads in the Rocky Mountains, the conspicuously loud chirping of

Figure 5.4 Birds featured heavily in *Descent of Man* as well. Darwin insisted that human races resulted from similar mate preferences to those that led to ornate feathers in these birds.

otherwise camouflaged crickets, big manes of male lions, and so on? Much as grandfather Erasmus Darwin speculated decades earlier, Charles posited *sexual* selection or, less technically, "mate choice." In that scheme, plain-appearing (to humans) females select males based on the pleasantness of their appearance or behavior. Women choose who to mate with, in other words, even when the choice seems to perpetuate some feature that gets in the way of survival. Indeed, mate choice can appear to lead to maladaptations – much like artificial selection for "cute" dogs can lead to maladaptations (e.g., floppy Golden Retriever ears breeding more ear mites, the sloping backs of German Shepherds making canine hip dysplasia more likely, or the scrunched face of a French bulldog leading to respiratory problems). Sexual selection, mate choice, could explain any feature from bird feathers to human skin tone.

With this sexual selection silver bullet, however, Darwin almost killed his entire project.

Survival was easy enough to grasp. Adaptation had been baked into the world of biology since Aristotle. Natural theologians loved to point to the complex fit between the trait and its environmental use as a reason to believe that God the Master Watchmaker painstakingly set it all in place. Now Darwin appeared to be shrugging all that off in a way that even *Origin of Species* and *Variation in Animals and Plants* avoided. It could be that a dramatic variety or a strange trait of an organism existed simply because it appealed to a female. And then to her daughter. And her granddaughter. And her great-granddaughter. Eventually, many generations later, what started off as a bird with slightly redder feathers became a cardinal, a slightly larger beak became a toucan, and slightly different color eyes or texture of hair became a different race of humanity. This was Darwin's Big Idea – not just a principle of divergence (a "push") culled by the struggle for existence (a "pull"), but an entirely new way of thinking about evolution (well, "new" if we discount that grandfather Erasmus also proposed a version of it). It was much, much broader than natural selection alone.

Even Alfred Russel Wallace couldn't stomach it when it came to humans. An ultra-adaptationist unable to handle the non-adaptive wiggle-room Darwin left even in his standard selection model, Wallace couldn't agree that female beetles and squids and worms made aesthetic choices akin to humans, deliberate choices he believed would be required for sexual selection to work.

The Expression of the Emotions in Man and Animals (1872)

Just as *Descent of Man* began as a "short essay" that popped out of his work in *Variation*, the third book of the "trilogy" emerged from thoughts too lumpy for *Descent*. Perhaps not surprisingly, *Expression* feels different than the earlier two books. Different, even, than *Origin of Species*. Unlike the previous books, Darwin did not include any novel process to account for evolutionary change. And neither pangenesis nor sexual selection play much of an explanatory role in *Expression*. Instead, Darwin targets the misgivings shared by Lyell long ago and reinforced in Wallace's growing interest in Spiritualism: To what degree are human emotions and other products of the mind products of common descent with modification? Are humans set apart somehow in their minds, if not in their bodies?

Darwin lays out *Expression* as an extended search for behavioral homologies – those traits conserved through multiple species and genera the way that a whale flipper, a human hand, and a bat wing deploy the same conserved carpal bones for different functions. Of course, he finds them. Repeatedly. From anger and fear to sadness and suffering, even to complex emotions like contempt, shame, modesty, and devotion, Darwin locates trends in emotional expression from very simple organisms all the way up through humankind.

A careful read reveals two themes in the work that, while Darwin does not bring undue attention to them, turn out to be very important for the future reception of Darwinism. The first is the specific medical language Darwin uses and the mechanical way he uses it. Drawing from French physiologists, he identifies specific muscles in the face, for instance, that produce expressions in humans and in familiar mammals like dogs and horses. This is as close to a direct homology of muscle to muscle as anyone had published outside of specialized medical journals to that point, and it reveals a cross-disciplinary specificity that Darwin's previous works don't exhibit. He doesn't say *how* those *depressores anguli oris* (muscles that turn the corners of the mouth down), to name one example, came to retract and expand the same way in completely different organisms, but they do. For Darwin, that is a strong bit of evidence in favor of common descent. He also pins behavioral expression to "nerve-force," states of the nervous system. Using language also applicable to machines in the Victorian era, Darwin highlights shifts of nerve-force, conduits in which it produces pressure, and explosive discharges of this energy in the bouncing of an excited child or the kicking of an expectant pony, for instance. Throughout, Darwin employs the language of physics and engineering – negative and positive charges, attractions and repulsions, inputs and outflows – to describe what many of us regard as strictly organic *feelings*.

We might feel even less sanguine about the second theme that appears in *Expression*. The biggest challenge in applying evolution to humanity is the intellectual gap that apparently exists between apes and humans, let alone that existing between us and other animals. After all, we study mice and dolphins and not the reverse (assuming Douglas Adams was writing fiction, I suppose). Darwin wanted to shrink that gap. So, not only did he dwell on the most complex of animal behaviors, he also highlighted the behavior of those humans that he regarded as the least complex.

To that end, Darwin began corresponding with the 29-year-old J. Crichton-Browne, director of the 1,000+ patient West Riding Pauper Lunatic Asylum at Wakefield in Yorkshire, England. Crichton-Browne gave Darwin unrestricted access to notes, records, and observations of his insane asylum patients. The mentally ill, believed Darwin, would show a humanity unvarnished by modern mores, the skein of civilization that might obscure the link between humans and animals torn away. The mentally ill were lower on the ladder of development, in other words, and would serve as kinds of "missing links."

Why the (ethically questionable) generosity by the Wakefield Asylum to give Darwin access – even sending 40 close-up photographs of the faces of patients in various emotional states, including real pain? Family connections. Crichton-Browne's father, the phrenologist W. A. F. Browne, studied with Robert Grant at Edinburgh and, like Grant, strongly endorsed Lamarckian and Erasmus Darwinian views. In fact, it was Browne who sponsored Darwin's membership in Edinburgh's Plinian Society all the way back in 1826. And it was Browne who offered an eerily similar argument in the 1820s, using the latest data from phrenology, that human and animal behavior shared the same evolutionary trajectory, using the same portions of the brain and even the same muscles to enact the expressions of emotions in both man and animals.

In other words, in *Expression of the Emotions*, Darwin used the observations of mentally disabled patients under the care of Crichton-Browne to support the point that Crichton-Browne's father, W. A. F. Browne, made a half-century earlier. As always, though, Darwin's exposition is methodical, careful not to offend. And his conclusion, foregone: evidence of descent with modification is present in human minds as well as bodies; the "lowest" humans show the most "animalistic" traits, and the reverse is true. The whole must be the descendants of one or a few parents.

Darwin's Darwinism

The roughly 16 years from his meeting with Huxley, Hooker, and Wollaston in April 1856 to the publication of *Expression* in 1872 proved incredibly productive. He didn't stop there. But none of the five(!) books he published in the last decade of his life rose to the same prominence as the major four published from 1859–72. *Insectivorous Plants*, published in 1875, *The Effects of Cross and Self Fertilisation in the Vegetable Kingdom*, published in 1876, *The Different Forms of Flowers on Plants of the Same Species*, published in 1877, *The Power of Movement in Plants*, published in 1880, and *The Formation of Vegetable Mould Through the Action of Worms*, published in 1881, reinforced the same processes he saw in every organism he studied. Variation, selection, common descent through variation and selection.

Darwin's Darwinism depicted a vital tendency to grow, spread, morph, and diversify countered by the problem of survival itself; environmental pressures leading to slight tweaks to eyes, hands, lungs, hair, leaves, roots, and, ultimately, teeth, bones, fruit, and seeds. Darwin's theory says nothing of eternal selfish genes but of flexible, gregarious gemmules – those bits floating in the liquid of the body, continually changed by environmental conditions and by use and disuse, coagulating at the time of reproduction, allowing a slightly varied line to continue in the face of almost unrelenting competition for resources. Other alterations to appearance and behavior emerge, not directly related to survival, perhaps, but important for females to choose their mates. All in all, it's a world in which the same organs, bones, and emotions appear in different organisms performing different functions. A world where there are no gaps from monad to man, just as grandfather Erasmus insisted many decades earlier.

In the background, even without him meaning to, Charles Darwin grew into a transformational figure, almost larger than life. His ideas of descent by agonizingly gradual modification swelled, taking on a prominence far greater than any published part of Darwin's ideas. They splashed over the riverbanks of biology and geology and into psychology and anthropology, history and philosophy and medicine and jurisprudence. Business leaders, military men, and politicians wallowed happily in survival-of-the-fittest pools. Muddy Darwinian floodwaters rose up even the marbled steps of religious temples.

He may have avoided the limelight like a snail avoids table salt. Yet Darwin's death and burial in 1882 alongside the great and the good under the flagstones in the floor of Westminster Abbey, an apple's throw from Sir Isaac Newton's ostentatious mausoleum, symbolized both the end of gentlemanly naturalism as frozen in the great natural history museums of the world and the cresting wave of imperial, combative, instrument-dependent, commercial, international Big Science.

6 Saint Charles's Place

The Good News finally snagged him. In late September 1881, he was near the end, bedridden, languishing in a soft purple robe, still able to read, though he always preferred to be read to. Lady Hope entered the drawing room at the top of the stairs quietly, respectfully, as the golden hour gently illuminated corn fields and English oak forests through his picturesque bay window. The faintest crown of white hair encircled his head in the late afternoon light; the rest was wizardly beard (Figure 6.1). Lady Hope, the well-known evangelist, was visiting the Darwins, and she approached the old scientist cautiously. But she needn't have. In his wrinkled hands he held the Bible, open to the New Testament Epistle of Hebrews. "The Royal Book," Darwin called it, serenely, mentioning a few favored passages.

Somewhat surprised, Lady Hope pressed him on his views on the book of Genesis. A "look of agony came over his face." Brows furrowed, pained even, he admitted that none of his claims in *Origin* or *Variation* or *Descent* or *Expression* or the other books had been definitive. They were just "queries, suggestions, wondering." All made by him in his youth, when his ideas were "unformed." It hurt that others had taken them, twisted them, made them into a "religion."

Perhaps knowing that his end was near, Charles Darwin mustered his energy, turned to Lady Hope and practically begged her to hold a Gospel service at his "summer house" in his garden, gesturing to a place just out his window. The summer house was big enough for one of her evangelizing meetings, for 30 people, perhaps. Once gathered, she should preach about "CHRIST JESUS" – Darwin was emphatic on this point. And then

Figure 6.1 Charles Darwin in 1881, around the time of visits by Lady Hope and Edward Aveling.

there should be singing; the new style of upbeat evangelist music Lady Hope played herself, on her own stringed instrument, none of that Church of England boring stuff. If she held her service at this same time, around 3 o'clock, she should know that he would be "joining in with the singing."

According to a 1915 Boston, Massachusetts, newsletter and several other American religious publications from around the same time, this is Darwin's "conversion story." Evangelistic tracts told and retold this story in the period between the First and Second World Wars. I heard an even more exciting version of it in the 1990s – that's the kind of staying power this story has. You can still find its dry bones jangling around in corners of the American evangelical internet even today.

How much of this account actually happened?

Behind that question is an even more important one: what was Darwin's take on Christianity in particular and religion more generally? Today's Darwinians, including well-known figures such as Richard Dawkins, go so far as to say that

Darwin endorsed atheism or made it possible to be a "fulfilled" atheist. In some sectors of evangelical atheism, you find Darwin portrayed as the bottle that held the solvent that broke the bonds of religion on the modern world.

His own writings, both public and private, and testimonies about him told by people who knew him best consistently reveal that advocacy for atheism is a misconception, at best. It might even be completely misleading. But it's tough to get at his views. To see the real religious convictions of Darwin, we must peer through a glass, darkly.

Darwin Never "Converted"

Recent accounts of Darwin's "deathbed conversion" read on social media or overheard at revival meetings further twist the already fuzzy testimony of Lady Hope into sheer fantasy. Even that original published story, which appeared during the First World War in the Boston, USA, evangelical newsletter, *The Watchman-Examiner*, departs from Lady Hope's actual reporting.

Lady Hope's own words didn't appear until years later, through her letter to a different editor, penned after the conclusion of the Great War when she was nearly bankrupt in New York City. It wasn't until 1919 or 1920 – some four decades after the events supposedly took place – that she recounted why she would have been in the area. Recently widowed, she was staying with friends in Kent, near the Darwins, in 1881. That summer, Lady Hope traveled unceasingly through the small villages of southeast England holding evangelical temperance revivals. In her 1919/1920 letter, she recalled being invited to visit the Darwin family at 3 pm (though she never revealed the date). When she entered Down House, she walked up the stairs to the drawing room and found Darwin reclining on a sofa beside the window. She repeated small details: the high ceilings, the way he held his hand out to her, his balding pate, the intense look of his eyes, and the fact that he was reading the book of Hebrews and called it "the Royal epistle," claiming that he never tired of it.

According to Lady Hope's 1919/20 letter, it was *Darwin* who lectured to *her* about "the King, the Saviour, the Intercessor, dying, living … with great animation on different parts of the subject." When she pivoted toward the book of Genesis – the juicy center of every version of this story – he naturally grimaced. He confessed his youth and ignorance, but insisted he wanted "the truth." (Note that he published the *Origin* at the age of 50, the *Variation–Descent–Expression* "trilogy" books in his late 50s and early 60s; claiming "youth" would have been ridiculous.) The climax of her story is what happened next: "He hesitated, as if he was quite overcome, and burst out with a louder voice, apparently in great displeasure, 'They went and made a religion out of it.' He sank back quite exhausted, after this outburst, and closed his eyes. Then we talked again quietly." Soon afterwards – either during this interaction or, Lady Hope mused, perhaps another later one – Darwin offered his "summer house," large enough to hold "thirty people" (coincidentally, 30 was the typical size of her Bible studies held inside the drawing rooms of other large country homes in Kent). She should hold a meeting out there somewhere on his property, to speak to his "servants and laborers and some tenants," to "sing some hymns, not the sad old drony ones, but those others." These "others" would be the hymns associated with American evangelist Ira D. Sankey, such as "Standing on the Promises" and "Tell Me the Old, Old Story," often scripted like military marches or popular songs of the day, with minimum choral complexity and maximum pathos. Sankey's singing accompanied Chicago revivalist Dwight L. Moody's emotional preaching on their wildly popular evangelistic tours of Britain in 1873–75 and 1881–84. Lady Hope and her troupe no doubt used the songs from Sankey's bestselling hymnals. Yet, as Lady Hope lamented, the Darwin garden-house meetings never took place. Someone unmentioned – someone other than the old "doctor" himself – didn't approve of her presence.

As one of Darwin's greatest biographers, James Moore, sleuthed decades ago, some of these details appear authentic. Her recollection of the room is correct: though in our contemporary museum version of the Darwins' Down House, that room is called Charles and Emma's bedroom at the top of the stairs from the ground floor, it was indeed an upper-storey drawing room in 1881. Darwin likely did recline on a sofa near and was able to see some of his rear garden from that window (Figure 6.2). He possibly did enjoy the sunsets – though the "golden hour" before sunset would still have been hours off in late September/ early October.

But many more details cannot be accurate. Though she mentions no one else with them, widowed, charismatic 39-year-old Lady Hope would not have sat

Figure 6.2   The view out the window where Darwin supposedly reclined in September 1881, when Lady Hope visited.

alone in that upper drawing room with Darwin – that just wasn't done in a respectable Victorian household. More centrally, he was several months away from death and still in the same not-quite-well health as always. He had never spoken about "renouncing" evolution to anyone (which, to be fair, Lady Hope herself never suggests). And even though he did study the Bible during his undergraduate days, there is no indication Charles Darwin would have been leafing through the epistle of Hebrews looking for inspiration before the final closing of his eyes, which didn't come for months anyways.

Born Elizabeth Reid Cotton in Australia in 1842 and raised in the Raj in India, "Lady Hope" married retired Admiral and First & Principal Naval Aide-de-Camp to the Queen, Sir James Hope, in 1877, when he was 69 years old, and she was 34. Sir James had been the heroic captain of the HMS *Racer*, a brig-sloop (like FitzRoy's *Beagle*) immortalized in paintings for miraculously surviving the catastrophic 1837 Gulf of Mexico Hurricane. Over the years, Captain Hope both rose in rank and increasingly turned toward evangelical expressions of Christianity that often motivated the temperance and anti-saloon movements. After his first wife died, he joined his newly minted Lady Hope. She made the connection even more explicit between faith and alcohol-abstinence and traveled around Britain preaching the gospel, leading Bible studies, playing inspirational American music, and agitating for temperance. When Admiral Hope died in 1881, Lady Hope intensified her teetotaling mission work, including visiting the homes of the wealthy across southeast England seeking donations to the cause.

No doubt it was for this reason that she arrived at the Darwins and, indeed, found herself in the drawing room with the elderly, but not yet dying, Charles Darwin. Emma almost certainly sat quietly nearby. Perhaps Joseph Parslow, the family butler who played with Charles in their new billiard room, helped arrange the visit. By all accounts, Parslow had some sort of a conversion. But to what? Was it belief in a Personal Jesus or living without alcohol that required the greater sacrifice for Parslow?

And what of the visit itself, which must have taken place between 29 September and 4 October 1881? The Darwin children (Francis, who compiled and edited his father's three-volume *Life and Letters*, perhaps most especially) vehemently denied such a meeting ever took place. Moreover, the elapsed time between the visit and Lady Hope's retelling of it for American religious periodicals weighs heavily against its veracity. Yet, as Moore, among others, insists, *something* must have occurred. Her advocacy for temperance no doubt struck a chord with the Darwins. She would be speaking to and singing with "servants and laborers and some tenants," not to the wealthy landowners themselves.

After examining the different accounts, the most likely version is that the Darwins welcomed Lady Hope politely and agreed that she could convert *the help* to cleaner, more honest living. It's quite possible that Lady Hope never questioned Darwin's own faith in the conversation(s).

But what if she had? How might Darwin have responded?

The Darwin I Never Knew

For Cambridge churchmen-in-training in the 1830s, like pre-*Beagle* Charles Darwin, William Paley's natural theology texts collectively painted a metaphysics in which God could still be kind, caring, and benevolent in a world that seemed anything but. Slavery of Africans and Indians and the brutal subjugation of Aboriginal peoples followed the Union Flag like a pestilential cloud wherever it flew over non-English lands. Institutional Christianity seemed flat-out powerless against this blatant injustice; sometimes the Church even encouraged it. Paley's pro-abolition Natural Theology offered a vision of the world that was originally built better than this. Evil, from that perspective, would prove to be superficial; *design* – with every organism fitting a proper place – the truer law of the world.

Drinking anti-slavery rhetoric with his mother's milk, Darwin saw through the eyes of Paley's Natural Theology a vision of a just and peaceful world still flourishing. A well-designed world where every strange feature of beetle, orchid, earthworm, or finch had its right place, hovering just behind a human world that couldn't perceive it or, when it could, perverted it.

By this reading, "survival of the fittest" meant that God so loved the world that He gave His only begotten Plan; that whosoever believes that God is an excellent Designer and does not work by caprice should not stay ignorant but have real understanding of the Natural Order. It's only with this deep belief that the world is working properly despite appearances that Darwin could draw on the grand illustration at the end of the *Origin*: the tangled bank, perhaps along the river of a rainforest Darwin traversed or just outside in his English garden, "clothed with many plants of many kinds, with birds singing on the bushes, with various insects flitting about, and with worms crawling through the damp earth . . . produced by laws acting around us." Only, it's not a relaxing place at all: "Thus, from the war of nature, from famine and death" come wispy orchids pollinated by jewel-colored hummingbirds, pods of humpbacks gliding through Pacific swells vacuuming plankton, lean wolves running down elk over snow-encrusted taiga, and pugs (okay, those take human breeding, but still). Rules once laid down, sure as gravity. They might look like destruction and misery to us, but underneath them stirs a plan to create "endless forms most beautiful and most wonderful."

It's mysterious. It can be awful and cruel. But it's not without *order*. Thus, Darwin did not see his theories as promoting atheism. So, not surprisingly, even if he refused to endorse Christian evangelicalism, he likewise refused to confess atheism, even though given several opportunities to do so both in public and privately.

Pressed by one of his admirers after the publication of *The Expression of the Emotions in Man and Animals*, published in 1872, to explain on what grounds he still believed in God, Darwin replied, "the impossibility of conceiving that this grand and wondrous universe, with our conscious selves, arose through chance, seems to me the chief argument for the existence of God." He scribbled these lines to a Dutch student who confessed unbelief and promised confidentiality – Darwin could easily have endorsed atheism in this letter but chose not to. The student, Nicolaas D. Doedes, gently criticized him for even hinting at orthodoxy in the *Origin* and even *Descent of Man*. But Darwin did not take the bait. He held out.

He offered to Doedes another reason to hang onto *something*. Billions of feet carrying weary heads to mountain shrines, gilt domes of cathedrals, ancient synagogues, ornate mosques. Trillions of prayers whispered by deadly earnest, weeping mouths – perhaps Darwin even recalled his own gut-wrenching prayers choked out in the dead of night while his beloved 10-year-old daughter, Annie, withered away and died from some unknown fever miles away from home in 1851. *So many believers believed so fervently*, individuals of every language and every culture in every country for century upon century of human history. Are we really willing to wipe that away so thoughtlessly? He knew this to be a weak argument, since lots of people believe lots of strange, unbelievable things. But, to Darwin, belief by proxy with the staggering communion of saints from time immemorial seemed better than unrelenting unbelief.

Actually (he seems to pivot even in this one letter), better to steer clear of reading theology into or out of the particulars of the natural world entirely. Puny human brains would never be able to adjudicate the existence of God once and for all. Instead of answering Doedes directly, Darwin offered, "man can do his duty" (DCP 8837). For God, Queen/King, and Country – even if you're not all that sure about the "God" part.

What duty? This seems a lame offering by Darwin.

On second thought, it's not a throwaway line at all. This line reveals a surprising amount about Darwin's faith.

Without meaning to, he had stumbled prematurely into a late-nineteenth-century arena wherein two powerful philosophers would later square off; W. K. Clifford and William James locked horns over exactly Darwin's position here. In 1879, Clifford defended withholding belief in possible religious humbug because it might lead to irrationalism: "The danger to society is not merely that it should believe wrong things, though that is great enough; but that it should become credulous and lose the habit of testing things and inquiring into them; for then it must sink back into savagery" (*Ethics of Belief*, 186). Darwin would no doubt have supported such a warning. But American philosopher William James formulated a rejoinder that would also have impressed Darwin: "a social organism [which naturalist Darwin knew well] is what it is because each member proceeds to his *own duty* with a trust that the other members will simultaneously do *theirs*. Wherever a desired result is achieved by the co-operation of many independent persons, its existence as a fact is a pure consequence of the precursive faith in one another of those immediately concerned" (*Will to Believe*, 24). *Acting as if* one believes and adheres to a set of moral principles derived from that belief can hold together societies, James thought.

Though we rarely discuss it, Darwin did this duty, this *acting as if* he believed, uncommonly well, trusting that society would work if everyone worked together for society. It was the surest path to peace and prosperity, "which is sure to follow you if you continue in your old upright prudent course," as he reminded his old *Beagle* servant, Syms Covington in 1843 (DCP 700). Supporting Lady Hope's temperance movement – though not necessarily her pathos-filled, sing-songy Americanized evangelicalism – could easily have fit this Darwinian system of belief.

No wonder the Darwins were such staunch supporters of their local church. St. Mary's, the Church of England cornerstone of Downe, acted as social glue. But that's not to say Charles and Emma Darwin acted according to purely instrumental motives either. Charles chose to bury his brother Erasmus there; Emma is buried in that churchyard, as is her sister Sara. Three of Darwin's children are also in that churchyard, marked by a large gravestone right near the entrance to St. Mary's. Emma and the children regularly attended services. They took the eucharist there. They faithfully subscribed to the Sunday School fund. The Darwins baptized their children in the Anglican faith. And, after Reverend John Brodie Innes (1817–94) took up the vicarage of Downe as a perpetual curate in 1846, the Darwins both supported the parish financially and through a considerable investment of their time. Charles Darwin headed Downe's Coal and Clothing Fund, the church affiliated charity, for over 20 years. He also helped Rev. Innes initiate, with his old professor Henslow's prompting, a Benefit Club called the Down [sic] Friendly Society, in 1850. The Darwins took a sustained financial interest. The club met annually at their house. Darwin acted as its treasurer – an uncompensated position that cost Darwin money without promising equal benefits to him or his family – until his death in 1882. (These mutual benefit societies were, and are, social insurance groups wherein members contribute subscriptions so that they could be cared for when ill or old, and to cover burial expenses.) When the Down Friendly Society threatened to close after Parliament passed the Friendly Societies Act of 1875 (apparently some members spread a conspiracy theory about the government nationalizing all benefit funds), Darwin argued strongly, and successfully, against dissolution. His motivation? "[T]he hope of doing some small good to my fellow Members. . ." (DCP 10853).

That faithfulness in church-civic matters made Darwin a long-time confidant of Rev. Innes. Partly for financial reasons, Innes was forced to move to Scotland, but continued to take an interest in Downe, visiting as often as he could. Darwin became a kind of proxy for the vicar, a gentleman pseudo-clergyman, the very life he and his family had envisioned for him while he was an undergraduate at Christ's College, Cambridge. The two corresponded about a range of issues, including Darwin's evolutionary theory. Innes settled on the canonical "two books" compromise and rested assured that Darwin was of the same mind. According to the "two books" metaphor – popular with Galileo, among others – Scripture is one book; science a different, comple-mentary book, both written to attain God's revelation to humanity. Or, as Innes put it, both science and religion reveal parallel truths written in opposite directions, as Hebrew and Arabic unroll in one direction and Latinate lan-guages in the opposite. As long as these non-overlapping magisteria

("NOMA," as coined by paleontologist Steven J. Gould in the twentieth century) recognize their different functions and their mutual truths, there can be no conflict: "How nicely things would go on if other folk were like Darwin and Brodie Innes!" the Reverend joked (DCP 11768).

It wasn't merely a ceasefire between otherwise oppositional partisans. As tension surrounding evolution, "materialism," modernism, socialism, and other seemingly anti-Church-isms mounted in the later-nineteenth century, Rev. Innes defended Darwin. At a Church congress, Innes assured his clerical superiors

> He is a man of the most perfect moral character, and his scrupulous regard for the strictest truth is above that of almost all men I know. I am quite persuaded that if on any morning he met with a fact which would clearly contradict one of his cherished theories he would not let the sun set before he made it known. I never saw a word in his writings which was an attack on Religion. (DCP 11768)

After Darwin's death, Innes continued to defend him:

> Becoming Vicar of Downe in 1846, we became friends, and so continued till his death. His conduct towards me and my family was one of unvary-ing kindness, and we repaid it by warm affection ... In all Parish matters he was an active assistant ... Having myself always endeavoured to be first a Churchman, it is impossible that such relations could have been maintained had Mr. Darwin been ... the opponent of religious truth which those supposed who did not know him... (Recollections, CUL-DAR112.B85–B92)

Odd behaviors by, and odder comments about, someone actively promoting atheism.

In fact, it was the church that pushed the Darwins away, not the reverse. In 1867, a new clergyman came to Downe, Samuel James O'Hara Horsman, and imme-diately began attempting to raise money for a new church pipe organ. Within a year, Horsman departed on a 3-week tour for a "change of air" (DCP 6223). Once on board a friend's luxury yacht, however, the clergyman didn't return. No pipe organ had yet appeared, though the reverend collected significant funds for it. Horsman also completely neglected the area's charity school, which fell into disrepair. He did not pay the Sunday School teachers, either. And he racked up a considerable bill at the local pub with no apparent intention to pay it. Reverend Innes and Darwin collaborated to remove Horsman, pay the Sunday School teachers and reimburse the pub, then repair the school. Indeed, the whole Darwin family chipped in, though Darwin feared it might be too late "for the character of the Church" (DCP 6242). He was right. Even with Horsman gone, some members of St. Mary's began to attend a local Baptist church-plant. Darwin found himself a possible material witness against Horsman in a lawsuit. And, when a new, pricklier priest, Rev. George Sketchley Ffinden, finally replaced Rev. Innes as Downe's vicar in 1871, Charles and Emma Darwin turned toward the ministry of local Christian evangelicals.

Wallis Nash, a barrister (trial lawyer), and his wife, Louisa A'Hmuty Nash, moved to Downe in 1873 and opened a Reading Room in the local school to promote temperance. Louisa welcomed a chapter of the popular "Band of Hope" teetotaler society launched by Baptists in the 1840s to the village (Figure 6.3). The Darwins supported it. Through their support, the two families became quite close during the 1870s. When, at the end of the decade, the Nashes moved to North America to help colonize the Oregon Territory, Darwin mourned: "You will both ever be a heavy loss here" (DCP 12446).

A new, young evangelist, James William Condell Fegan, soon arrived in Downe with his parents, eager to continue the work of the Nashes. Darwin supported him as well. Fegan worked with orphans and other children of the poor and, in the summer of 1880, brought a gaggle of boys around to sing hymns for the Darwins in a process Fegan called "camping out." Charles gave each of them sixpence, to raucous cheers. Then, in February 1880, Fegan asked to take over administration of the old Band of Hope reading room and rename it the "Gospel Room." Darwin encouraged him:

> You have far more right to it than we have, for your services have done more for the village in a few months than all our efforts for many years. We have never been able to reclaim a drunkard, but through your services I do not know that there is a drunkard left in the village.

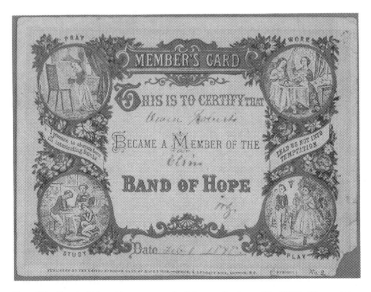

Figure 6.3 Band of Hope Temperance Society member's card (c. 1870). The society motto was "Thy Word Is Truth" emblazoned across an open Bible.

> Now may I have the pleasure of handing the Reading Room over to you? Perhaps, if we should want it some night for a special purpose, you will be good enough to let us use it (DCP 12879).

It remained Fegan's Gospel Room for over 50 years. The otherwise regimented Darwin family rearranged their mealtimes to allow anyone in their household to attend Fegan's evangelistic services. According to Fegan, preaching and Bible study at the Gospel Room brought Parslow the butler and Mrs. Sales the Darwin housekeeper "into the light." Emma rejoiced even more that Fegan converted the town blacksmith, a "notable old drunkard," in 1881.

Perhaps most surprising for those of us accustomed to viewing Charles Darwin as some kind of "Devil's Chaplain" is his longstanding support for the work of Christian missionaries in South America. Many years after they sailed together, teetotaler Admiral Sir B. James Sulivan (he was second lieutenant on Darwin's

Beagle voyage) convinced Darwin to support the South American Mission to the Fuegians at Ushuaia in the Beagle Channel, Patagonia. Darwin donated to Waite Hockin Stirling, Superintendent missionary of Tierra del Fuego and Bishop of the Falklands, and his partner, Thomas Bridges, in 1869, and continued to correspond with Admiral Sulivan regarding the mission for many years. Though a "civilizing" mission, Bishop Stirling and the other missionaries unabashedly spread Christianity to the Fuegians, as Darwin learned through the copies of *South American Missionary Magazine* Admiral Sulivan sent him. Sulivan also convinced Darwin to subscribe "as long as I live" to mission work assisting the orphaned grandson of Jemmy Button/Orundellico and to "Christianize" the grandson's name from his native Cooshaipunjiz to "Jemmy FitzRoy Button" (DCP 11481). Far from this bothering Darwin, he stood impressed with the mission work: "It is most wonderful, & shames me, as I always prophecied utter failure. It is a grand success" (DCP 7256).

On balance, then, the visit of Lady Hope to the Darwin home in the autumn of 1881 seems not at all improbable. Whatever his private thoughts, the *Origin of Species* author may even have found her work as a teetotaler and evangelist of the working class happily aligned with his own hopes that everyone in civilized society could *do their duty*.

The Brush with Atheism

So pronounced was Darwin's commitment to order even when the world appeared disorderly that atheists, when Darwin met them in the flesh, got a surprising brush-off.

Just before lunch on Wednesday, 28 September 1881 (days before Lady Hope's visit), Edward Bibbens Aveling, a lecturer from University College and anatomist at London Hospital, appeared at Down House. His hair parted severely over his sharp face – a "'little lizard of a man'," scorned one acquaintance. Beside him bowed Dr. F. K. C. Ludwig Büchner from Tübingen, Germany, eyes always in a haughty half-squint. They had given very little advance warning of their arrival. Emma, at least, disapproved, though she and the staff greeted them courteously, as always. In the future, Aveling would marry Karl Marx's daughter, Eleanor, drive her to kill herself, and inherit the Marx money. At the time he visited the Darwins, however, his fortunes appeared bleaker. He had lost his University College lectureship when, according to him, his advocacy for atheism – though really it was his looseness with money, women, and the money of those women – became too much for the London medical establishment. He maintained an income as an educator at the National Secular Society's science museum on Old Street, London, and was by all accounts an excellent speaker. And he wrote expository books, like *The Student's Darwin*, which he published a few months earlier with a laudatory dedication to the elderly naturalist, who really didn't want the dedication (Aveling wrote a similar primer for Marx years later). Darwin read the complimentary copy Aveling sent him in the summer of 1881 and likely disagreed with some of it, especially the anti-religious extrapolation (DCP 13287). He had already made it clear to Aveling that he avoided writing on religion; and whatever his private support for secular science, he would not directly attack Christianity (DCP 12757).

Emma's real concern, however, was Büchner, the German. The former path-ologist had founded the *Deutsche Freidenkerbund* (German Free-thinkers' Society) just a few months earlier in 1881 and was touring Britain, speaking at international secularist meetings. His books made him a star of the move-ment. *Kraft und Stoff* (*Force and Matter*, published in 1855), *Natur und Geist* (*Nature and Spirit*, published in 1857), and *Die Stellung des Menschen in der Natur* (*Man's Place in Nature*, published in 1869) hammered at Christianity, denounced any link between government and religion, mocked the simple-mindedness of believers in anything but atoms and void, and saw justification in biology for human social and economic competition, even if it led to inequality and misery for the lower classes. Much of his recent inspiration, he proclaimed, came from the works of Charles Darwin.

Lunch was as tense as Emma feared. According to her, Büchner spoke a great deal in German, excluding the other very English guests at the table, and he let fly his confrontational atheism. After the meal, Aveling, Büchner, Charles Darwin, and his recently widowed son, Francis, retired to the study for a smoke. The Darwin men seemed to find little joy in the meeting. Aveling and Büchner wanted a definitive anti-religious claim from Darwin; he just couldn't give one – he didn't think humans could sort out metaphysical issues like this. "There is about as much use in trying to illuminate the subject with the light of intellect as there would be in trying to illuminate the midnight sky with a candle," as one of his friends eloquently put it in 1878 (DCP 11779). Borrowing a term from Huxley, Darwin grasped for that deep feeling of unknowing: "agnostic." But Aveling scoffed. That was just weak atheism. Darwin disagreed: it was an act of faith to believe in *nothing*, as well. Better to just stay silent, steady on.

Büchner left unimpressed. This Darwin would not advocate a revolution that required the religious foundations of Western Society to be torn-up. Though they later tried to dress up the interaction, Charles Darwin was not the great champion of atheism the secularists wished him to be.

Even his young protégé, George J. Romanes (1848–94) – one of the few stalwarts supporting Darwin's pangenesis in the years after *Variation in Animals and Plants Under Domestication*, published in 1868 – couldn't shake an admission of atheism loose from Darwin. In *A Candid Examination of Theism*, published in 1878, which Romanes published anonymously, he insisted that Büchner was right, that everything was just matter floating in nothingness, that even human thought was just the outgrowth of impersonal laws, that atheistic materialism could be the only rational metaphysics from a scientific point of view. Darwin demurred. A theologian might easily say that God set up matter to act according to natural laws that would, in the end, result in human cognition and all the other wonders of the Universe. Paley might have incorporated this into Natural Theology as evidence of divine design – no atheistic philosophy needed.

In truth, authors closer in time than Paley made precisely this argument from design. Josiah Parsons Cooke, a Harvard University chemist, replied to the secularism accreting around Darwin's *Origin of Species* in his own *Religion and Chemistry; or, Proofs of God's Plan in the Atmosphere and its Elements*, published in 1864. Properties exhibited by chemicals found everywhere signaled a kind of fine-tuning to the Universe, argued Cooke. And given that chemicals are the preconditions for life to exist in the first place, one needn't be shaken by any latent atheism in Darwinian biology. God so loved the world that He designed hydrogen and oxygen, explosive gases, to forge water, the life-giver. With a wink toward Erasmus and Charles Darwin, Lamarck, and the whole prior century of transmutationism, Cooke declared that "no theories of organic development" could wipe away the fingerprints of the Creator revealed by chemistry itself.

Decisively Undecided

Those that find religious belief and personal actions deeply disconnected (i.e., those who regard doctrinal statements or "sinners' prayers" as more indicative of belief than one's long trajectory of behaviors – or those who deny you can tell a tree by its fruit) might be frustrated that Darwin proved extremely reticent to come clean on his innermost beliefs. They look at Darwin's actions as spelled out above and ignore the writings of both William James and St. James that faith and works travel in tandem. Perhaps they expect Darwin to come right out like his younger followers T. H. Huxley or John Tyndall or Ernst Haeckel and tell us how precisely to arrange our own belief systems given his scientific work. But that's not Charles Darwin.

He did drop hints, however. In 1874, American historian John Fiske published *Outlines of Cosmic Philosophy*, an attempt at synthesizing evolutionary science and faith. Darwin wrote, impressed: "here & there I had arrived from my own crude thoughts at some of the same conclusions with you" (DCP 9749). Fiske had advocated "cosmism," god as impersonal but involved somehow, a force directing the Universe from the inside: "a Power which is beyond Humanity, and upon which Humanity depends," wrote Fiske. God, in Fiske's rendering, might grow, change, develop, progress – but not like a *being* to be anthropomorphized. God would be a stretching of existence itself to make room for intelligent beings. That sort of god could be discovered in the laws of the Universe, but should still engender awe in lowly mortals like ourselves. Fiske's "sense of sublimity" resonated with Darwin.

He occasionally admitted something closer to conventional faith. "Fluctuations," he called them. "This follows," he confessed in his autobiography, "from the . . . impossibility of conceiving this immense and wonderful universe . . . as the result of blind chance or necessity." From this reflection, Darwin jumped to belief in "a First Cause having an intelligent mind in some degree analogous to that of man; and I deserve to be called a Theist." Ironically, as his son, Francis, recorded in an essay after his father's death, Darwin felt this most strongly when he wrote *Origin of Species* in 1859 – for which so many would condemn him as an enemy of faith.

For his part, Harvard botanist Asa Gray, close confidant of Darwin's and a devout Presbyterian, stoutly defended Darwin as holding perfectly respectable ideas. "You see *what uphill work* I do in making a theist of you, 'of good and reputable standing,'" Gray joked in 1874 (DCP 9492). But he had some material to work with. As Darwin mused after reading Gray's synthesis of evolution and Christianity – essays that Darwin called "*far* the best Theistic essays I ever read" – "I have a feeling that the existence of the multitude of Stars & the motion of the planetary system &c are equally good with living beings to prove a First Cause; yet if there were no living things, there could hardly be design. – But I well know that I am muddled-headed on this subject" (DCP 2930). Some version of theism did seem to comport with some version of evolution. Several high-profile late-nineteenth-century theologians, including B. B. Warfield, author of the Biblical inerrancy doctrine, clicked some version of evolution and Christianity together.

Darwin couldn't commit to this either, however. Pain and suffering, which his theories now baked into the natural order, proved far too daunting – especially after, in April 1851, he watched the apple of his eye, his innocent, joyous, intelligent 10-year-old daughter Annie (Figure 6.4) dissolve into cold death right before his tear-filled eyes.

Moreover, even when moved to some sort of belief, he wondered just how strongly we should regard any transitory emotional current, any feeling of awe, no matter how profound. Can we trust something as ephemeral and liable to mistakes as the "muddled-headed" human? Darwin committed to the unknowing, a believer in the unreliability of both rational orthodoxy and sentiment. He wanted enough evidence in the same direction to induce the correct answer. But when it came to religious conviction, evidence in the same direction proved too elusive. "The strongest argument for the existence of God, as it seems to me," wrote Darwin in 1878, "is the instinct or intuition which we all (as I suppose) feel that there must have been an intelligent beginner of the Universe; but then comes the doubt and difficulty whether such intuitions are trustworthy. . . . I am forced to leave the problem insoluble" (DCP 11416). In the end, the best he could do was throw up his hands: "I cannot pretend to throw the least light on such abstruse problems. The mystery of the beginning of all things is insoluble by us, and I for one must be content to remain an Agnostic" – not "un-belief" but "a-belief," as he once distinguished it.

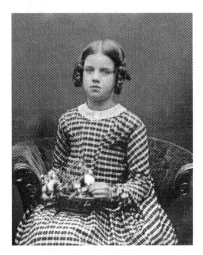

Figure 6.4 Annie Darwin in 1849 at eight years old. She would die less than two years later of an unknown illness.

In the meantime, in the absence of satisfactory answers, you still have a role in this world, a *duty*. Just grit your teeth and do it.

What Child Is This?

Ultimately, Darwinism only looks godless if you want it to look godless, if you expect your God to make only pleasant plans without mass death and suffering, without trillions of fertilized seeds that never sprout, trillions of eggs that never grow to adulthood, trillions of embryos aborted, trillions of parasites making trillions of organisms live shorter, more painful lives. Jesus of Nazareth proclaimed that God tracked the death of sparrows and valued us more highly than many of them. Biology presents us with a Golgotha of dead sparrows. Geology stretches that pile of sparrows back in time and joins them to even larger piles, uncountable numbers of extinct species of every imaginable shape and size. In other words, the real conflict here was never between

Darwin and religion; it's between the day-to-day mindless waste apparent in nature and the self-importance of *us*.

This is why so many so-called Scientific Creationists feel the need to strike out not only at what they conceive to be Darwin's theory but at the age of Earth accepted by geologists.

At the Creation Museum in Kentucky in the United States, animatronic dinosaurs wobble and roar (sounding more like modern mammals than ancient lizards) next to mannequins of ancient humans, with whom they are supposed to have shared the same air, water, and food – but not ancestry. One needn't dispute *Darwin* here but *Lyell*, the extreme age revealed by the geologic column. Why? Because if there's not enough time, natural selection can at most slide grasshoppers into other locusts, pines into fir trees, and so on. *Micro*evolution, the Creationists call it, and they're mostly fine with it. That lack of time wouldn't allow a gradual transmutation from, say, a primitive, four-footed synapsid into, on the one hand, finches Darwin could swat with his floppy hat on the Galápagos and, on the other, my favorite human-sized, sickle-toed death-raptor, *Deinonychus antirrhopus*. That's *macro*evolution. That's not palatable.

The reason is transparent enough if you stop to consider real living things (which is why most scientifically trained Creationists are engineers, accountants, lawyers, astronomers, physicists, mathematicians, or chemists – those who do not study organisms in situ – rather than biologists or geologists). On a geologic timescale, we humans are recent interlopers on a planetary graveyard. On an old Earth, by sheer volume, humans matter much *less* than sparrows, much less than vast numbers of trilobites or any of the other once incredibly fecund things that no longer creep, swim, or fly on the face of Earth. On an old Earth, Christ waited until absurdly late in a senselessly bloody game to bother redeeming anything. On a 6,000-year-young Earth, humans have always mattered; Jesus showed up roughly halfway through the story to make all the waste (of which there is far less if Earth isn't that old) cosmically matter.

Darwinism, predicated on that antique Earth, *can* accommodate God. But it's a God utterly alien to the human-centric, love-filled narratives that anchor many religions. Maybe even repugnant. It's a God that strolled on verdant plains alongside allosaurs as they shredded the desperately hooting babies of long-necked sauropods, grinned as plump caterpillars convulsed and burst open to reveal parasitic wasps laid inside them while still alive, nodded approvingly as volcanic outpourings nearly choked out life wholesale at the end of the Permian, and gazed out over a planet filled with even more gruesome scenes than these for millions upon millions of years, still pronouncing with a satisfied sigh that It Was Good.

7 The Struggle Is Real

Mythology piles like snowdrifts around many aspects of the history of Darwinism. Take, for example, the 1925 case of *The Board of Education of the State of Tennessee v. John T. Scopes*, better known as the Scopes "Monkey Trial." At the center of that media-generated maelstrom stood two heavyweight pugilists from the American heartland: in one corner Clarence Darrow, the trial attorney from rural Ohio, then Chicago; in the other, "The Great Commoner" William Jennings Bryan, one of the most accomplished orators of his day and a repeated Democratic candidate for President of the United States. Today our popular myths too often portray Darrow as a legal sharpshooter and Bryan as a bloated buffoon. But of the two of them, Bryan understood early-twentieth-century Darwinism far better. That's not to say Bryan understood *Darwin* or the scientific contributions to evolutionary theory. But Bryan sensed the sociocultural movement that trailed Darwin, borrowing a great deal from other sources, descending with modification, as it were, into something that looked substantially different than what Darwin himself wrote or intended. Darwin-*ism*.

For all their scientific prowess and public renown, there is no comparable Lyell-ism, Faraday-ism, Einstein-ism, Curie-ism, Hawking-ism, or deGrasse-Tyson-ism. So, there must be something even more powerful than scientific ideas alone caught in the net of this *ism* attached to Darwin. And whatever the term meant, it's fair to say that Darwinism frightened Bryan.

But why? And what does that have to do with the rest of Darwin's legacy?

We throw around terms like "Social Darwinism" and, what we (wrongly) presume is its scientific twin, "eugenics," without exploring what the work of Charles Darwin had to do with them. It's almost as if we believe someone merely had to utter "survival of the fittest" to uncork a malevolent force of exploitation in nineteenth-century European colonization that transmogrified into systematic genocide in the Central and Eastern European countryside by the 1940s. Some authors go so far as to say that, minus Darwin, there could never have been a Hitler; that the demon seed planted by the unassuming English barnacle-, orchid-, and earthworm-naturalist bloomed into Dachau, Treblinka, and Auschwitz-Birkenau.

I wrote that last sentence that way to emphasize just how clearly *this is a misconception*: millions of people participated in the injustices of the nineteenth and twentieth centuries that had never heard of, let alone read, the work of Darwin. American, Japanese, British, French, German, and other militaries constructed concentration camps in their colonies from the Philippines to Namibia that killed thousands through neglect, starvation, and disease, in addition to violence. Darwin doesn't figure in those stories in the slightest.

But that doesn't completely foreclose the concern. William Jennings Bryan saw *something* that so bothered him that he would make a whole campaign of discrediting Darwinism, including the infamous showdown with Clarence Darrow in the stifling heat outside the Dayton, Tennessee courthouse that contributed to Bryan's death just a few days later. Perhaps it's easy enough to dismiss him as just another blowhard politician, just as it's easy to dismiss the existence of any line from Darwin to death camps – that is, as long as you remain blissfully unfamiliar with the uncomfortable details of the social, political, and economic history surrounding Darwinism.

The *Allmacht*

Dr. Vernon Kellogg – no relation to Michigan's breakfast cereal magnates – was employed as a lowly insect biologist at Leland Stanford Jr. Teaching College, now known as Stanford University, in Palo Alto, California. But in 1915–16, he sat at the highest tables in Belgium, alongside the most powerful men in the world. It was the Great War. Kellogg sat with Germans in the

Grosses Hauptquartier (Central Headquarters) on the River Meuse/Maas. He was there because President Woodrow Wilson had so far kept the United States nominally neutral in the conflict. Wilson and Congress appointed American mining manager and alumnus of Stanford's Teaching College, Herbert Hoover, to administer humanitarian aid to Belgium and occupied France as the Commission for Relief in Belgium (CRB). By the end of the war, the CRB heroically delivered almost 5.7 million tons of flour, sugar, and other foodstuffs to keep Belgians and French from starving. Under the non-combatant CRB flag, Hoover tapped Dr. Vernon Kellogg, entomologist and ardent pacifist, to head the CRB for occupied northern France. From that post, Kellogg spoke with gray-clad German generals and chiefs of staff, and encountered even Kaiser Wilhelm himself. His nightly meetings with German officers and intellectuals at HQ, far from softening him, actually instigated his transition from pacifist humanitarian to aggressive pro-war, anti-German advocate.

Headquarters Nights, published in 1917, details Kellogg's conversion, which took place as the American biologist interacted more with the German planners of the war, men who left jobs as attorneys, surveyors, engineers, chemists, bankers, and, indeed, university biologists to don the gray uniform. Finally, Kellogg believed he understood their *Weltanschauung*, their worldview, in their own words. "Neo-Darwinism," Kellogg flatly reported. "The creed of the *Allmacht* [supreme power or principle] of natural selection based on violent and fatal competitive struggle is the gospel of the German intellectuals; all else is illusion and anathema." This confession shocked and disheartened Kellogg, who was himself a stalwart defender of evolution. In fact, in 1908 Kellogg had authored *Darwinism To-Day*, a robust defense of his (Kellogg's) laboratory work, which patched up the holes in Darwin's theory already apparent by the early twentieth century. Kellogg, in other words, was a neo-Darwinist.

The problem was how the Germans interpreted Darwin. For Alfred Russel Wallace (see Chapter 4 above) the *Allmacht* of natural selection drove him to setting humanity apart from the rest of the natural world. In German-speaking circles, however, Ernst Haeckel, one of the German-speaking materialist devotees of Darwin in the late nineteenth century, carried more weight than Wallace. Men like Ludwig Büchner, whom Emma found so objectionable when he visited in 1881. Men with political agendas. Men who believed that *Allmacht* applied to humans in all of our racial diversity, too.

In that truly global struggle for existence, a struggle which the German commanders believed applied to nation-states and human races, even if other Darwinians disagreed, only one group, one *Kultur*, would grow and develop enough to dominate or eliminate the others. Darwin, they insisted, had demonstrated this natural law for all living things. When applied to humans, later historians and critics labeled this attitude "Social Darwinism" – the strong oppressing the weak because nature told them to. Social Darwinism meant, Kellogg learned, that German officers opposed mercy or "human softheartedness." Not because they themselves were mean or cruel. But because survival of the fittest truly is an iron law. A nation could no more go against it than it could defy the laws of thermodynamics. What room could there be for supporting the weak, especially if they derived from a lesser people?

Kellogg's discovery of German Social Darwinism reported in *Headquarters Nights* didn't change his own scientific commitments. But it tipped the scales against Darwinism for William Jennings Bryan. After reading Kellogg's account, the Great Commoner feared that otherwise intelligent, sophisticated, cultured people could dredge up in the work of Darwin justification for war and the possible extermination of other groups. It motivated him to begin speaking out against the science itself. Well, "science falsely so-called," Bryan corrected – it didn't follow Newton's model of good science (see Chapter 5, above).

Even when the Great War ended with Germany's defeat and the breakup of the Austro-Hungarian Empire, Bryan continued to sound the alarm. We know that Darwin may have wrung his hands over the potential damage to the social order brought by a strong endorsement of materialism or atheism (see Chapter 6, above). But Bryan asserted that Darwinism *was* materialism or atheism. Maybe he was a little disingenuous here; maybe he knew the case wasn't so clear that Darwinism was atheism. Yet Bryan feared Darwinism meant the struggle for existence would pit the socially, politically, and economically strong in the USA – in other words, the glitterati, urbanites, businessmen, amoral Republican policy makers in Washington, and cynical East Coast journalists (perhaps all those people with large Instagram followings today) – against his people: the masses, those who went to church more than school, Southerners, Westerners, populists, Democrats. Far more than a mere scientific theory, Bryan proclaimed, Darwinism leached meaning, leached altruism, leached morality. Perhaps, as he believed it did in Germany, it would turn America into a dog-eat-dog nation (as if it wasn't always thus).

To be clear, Charles "Man can do his duty" Darwin never said these things. Most of the sentiments long pre-date him. We can see "man is wolf to man" antecedents all the way through human history. Two centuries before Darwin, philosopher Thomas Hobbes famously appealed to the necessity of a sovereign ruler to keep humanity from being in an all-versus-all war. The concept of a global struggle for existence in the animal world long pre-dated Darwin as well. Other well-known authors even popularized the notion that human races compete, with the strongest dominating or extinguishing the others, before Darwin penned his evolution books. For instance, in *Types of Mankind*, published in 1854 – a lavishly illustrated, scientifically robust (for the time), collaborative project between distinguished anatomists, linguists, archaeologists, and geologists that out-sold Darwin's *Descent of Man* for most of the nineteenth century – Josiah Nott, George Gliddon and the other authors asserted that only conquest and colonization could raise the overall level of humanity:

> [H]uman progress has arisen mainly from the war of races. All the great impulses that have been given to it from time to time have been the results of conquests and colonizations. Certain races would be stationary and barbarous forever, were it not for the introduction of new blood . . . and some of the lowest types are hopelessly beyond the reach even of these salutary stimulants to melioration (53).

Clearly, German officers had no need to reference Darwin to forge the legendary war between races in Europe; the sentiment could be found all over. Moreover, as they demonstrated in their southwest African geno-cide, where the Kaiser's *Schutztruppe* murdered perhaps 80,000 Herero and Nama men, women, and children in the lead up to the First World War, the German military required no justification in biology to commit atrocities. As many American and European colonizers have demon-strated over the last few centuries, death camps don't require Darwinism.

Survival of the . . .?

It's not just that Darwin didn't explicitly endorse the cruelty on display here. Much like his other family members who stood for the abolition of the slave trade, Darwin actively expressed disgust at cruelty to enslaved humans. He displayed this family aversion at his first confrontation with the totalizing enslavement system in Salvador, Brazil, in 1833, which he wrote about passionately in the 1845 second edition of his HMS *Beagle Journal of Researches*:

> Near Rio de Janeiro I lived opposite to an old lady, who kept screws to crush the fingers of her female slaves. I have stayed in a house where a young household mulatto, daily and hourly, was reviled, beaten, and persecuted enough to break the spirit of the lowest animal .... I was present when a kind-hearted man was on the point of separating forever the men, women, and children of a large number of families who had long lived together. I will not even allude to the many heart-sickening atrocities which I authentically heard of . . . nor would I have mentioned the above revolting details, had I not met with several people . . . [who] speak of slavery as a tolerable evil. (333)

Without question, young Darwin continued to carry this Wedgwood-family torch against the cruelty and violence needed to uphold the intercontinental economic institution of human enslavement. When Britain moved to finally eradicate it in 1833 (having done so only incompletely just before his birth), Darwin wrote to his sister, Catherine:

> I have watched how steadily the general feeling, as shown at elections, has been rising against Slavery. – What a proud thing for England, if she is the first Europæan nation which utterly abolishes it. – I was told before leaving England, that after living in Slave countries: all my opinions would be altered; the only alteration I am aware of is forming a much higher estimate of the Negros character. – it is impossible to see a negro & not feel kindly towards him; such cheerful, open honest expressions & such fine muscular bodies; I never saw any of the diminutive Portuguese with their murderous countenances, without almost wishing for Brazil to follow the example of Hayti; & considering the enormous healthy looking

black population, it will be wonderful if at some future day it does not take place. (DCP 206)

In this now widely quoted passage, Darwin reveals both his cordial feelings for the enslaved man's physique and cheerfulness and contrasts that with dark descriptions of the European enslavers. We may now read this sentiment as a cringe-worthy repetition of the "happy slave" anecdote ever after baked into the pernicious American Lost Cause myth. But importantly, this letter suggests that Darwin had no particular antipathy toward the personages of the enslaved. Cruelty against them he abhorred without hesitation.

In his published works, Darwin repeatedly argued for the shared ancestry of all organisms, including humans. This is probably the most well-known of Darwin's contributions, even if others, including his own grandfather, expressed it earlier. So convinced was Darwin of this singular point that he ended the final book in his evolutionary trilogy, *Expression of the Emotions in Man and Animals*, with another appeal to common humanity as shown through shared expressions of joy and sorrow. Under the skin, we are all basically the same, Darwin seemed to say. In fact, we are all a lot like dogs, orangutans, chimpanzees, gibbons, cats, dogs, horses, and birds, too (see Chapter 5, above).

We can place all this on the positive side of Darwin's reputational ledger. Unfortunately, there's plenty to put on the negative side, especially in his private correspondence. Truly, plenty of Darwin's own words leave open the possibility for an extreme "Social Darwinist" interpretation of the "survival of the fittest" language that Darwin adopted from sociologist Herbert Spencer at the behest of Alfred Russel Wallace in later editions of his *Origin of Species*.

For instance, in the weeks preceding the publication of *Origin of Species* in November 1859, Darwin sent printer's proofs to Charles Lyell. The two then corresponded regarding Lyell's objections to Darwin's theory. Darwin identified what he thought to be the core of their dispute. Though it is true that Darwin explicitly avoided discussing humans in *Origin*, as he had explained to Wallace (see Chapter 4), Darwin thought an example using human race would drive home the message of the *Origin* to Lyell:

As far as I understand your remarks & illustrations, you doubt the possibility of gradations of intellectual powers. Now it seems to me looking to existing animals alone, that we have a very fine gradation in the intellectual powers of the Vertebrata, with one rather wide gap (not half so wide as in many cases of corporeal structure) between say a Hottentot & an Ourang, even if civilised as much mentally as dog has been from wolf.

I suppose that you do not doubt that the intellectual powers are as important for the welfare of each being, as corporeal structure: if so, I can see no difficulty in the most intellectual individuals of a species being continually selected; & the intellect of the new species thus improved, aided probably by effects of inherited mental exercise. I look at this process as now going on with the races of man; the less intellectual races being exterminated. (DCP 2503)

Evident in this passage is their shared prejudice that native southern African Khoikoi and Khoi-san ["Hottentots"] and orangutans remained distinguishable by a wide-but-not-*that*-wide intellectual gap between them, "even if civilized." The harsh implications of this comment are hard to stomach. But in my mind, they're overshadowed by Darwin's admission that non-whites were just then being exterminated across the globe because they were "less intellectual."

Perhaps this was only a single letter, however; not indicative of a larger pattern, necessarily.

Sadly, there was more. Much more. Darwin and Lyell continued their friendly sparring uninterrupted through 1860, as they had for many years, returning repeatedly to issues of human race. In September 1860, they both referred to the much grander and more impressive *Types of Mankind* collaboration (quoted above), in which the authors defended separate origins for each human race (polygenism). Darwin found the scientific arguments for polygenism weaker than did Lyell but did not challenge the book's open bigotry and support for a so-called war between human races.

All the races of man are so infinitely closer together than to any ape, that (as in case of descent of all mammals from one progenitor) I sh^d look at all races of man as having certainly descended from single parent. – I should look at it as probable that the races of man were less numerous & less divergent formerly than now; unless indeed some lower & more aberrant race, even than the Hottentot, has become extinct. … Agassiz & Co. [authors of *Types of Mankind*] think the Negro & Caucasian are now distinct species; & it is a mere vain discussion, whether when they were rather less distinct they would, on this standard of specific value, deserve to be called species. …

I agree with your answer which you give to yourself on these points; & the simile of man now keeping down any new man which might be developed strikes me as good & new. White man is "improving off the face of the earth" even races nearly his equals. (DCP 2925)

Here again, you might find it as jarring as I do to read Darwin's cavalier references to white supremacy and genocide as supportive of his overall theory in *Origin of Species*.

It wasn't just correspondence with his old mentor in geology, Lyell. Similar themes appear in correspondence to some of Darwin's other prominent correspondents. In 1862, Rev. Charles Kingsley, social reformer, historian, and popular author of *The Water-Babies*, speculated that legends of dwarves, fairies, and other mythical humanoids emerged from real "missing links" between humans and apes. "That they should have died out, by simple natural selection, before the superior white race, you & I can easily understand," wrote Kingsley (DCP 3426).

Darwin expressed skepticism about the roots of these legends but, just as he had with Lyell, agreed with Kingsley about whites exterminating all other races: "In 500 years how the Anglo-saxon race will have spread & exterminated whole nations. …" But it's Darwin's next sentiment that seems still more upsetting: "in consequence how much the Human race, viewed as a unit, will have risen in rank" (DCP 3439).

To different correspondents many years later, Darwin reaffirmed this view: "When I look to the future of the world hardly any event seems to me of such great importance as the settling of Australia, New Zealand, &c by the so called Anglo Saxons…" (DCP 12158). And again:

Lastly I could show fight on natural selection having done and doing more for the progress of civilisation than you seem inclined to admit.

Remember what risk the nations of Europe ran, not so many centuries ago of being overwhelmed by the Turks, and how ridiculous such an idea now is in [*sic*] more civilised so-called Caucasian races have beaten the Turkish hollow in the struggle for existence. Looking to the world at no very distant date, what an endless number of the lower races will have been eliminated by the higher civilised races throughout the world. (DCP 13230)

Overall, these letters show, shockingly, that the German officers who bragged to Kellogg over dinner at HQ in 1916 about exterminating weaker races *would* have found a portion of their justification in these exchanges. If they'd known about them – which, of course, they did not.

What about Darwin's publications?

Though less explicit, they remain problematic. Begin with *Origin of Species*. As he mentioned to Wallace, Darwin avoided references to humans in the work. Yet he anchored his account around inter*racial* variation and competition. We often discuss competition between individuals in pursuit of resources and mates as the heart of Darwin's theory (so-called 'survival of the fittest'), and this is true. But as discussed in Chapter 5, above, the rarely acknowledged subtitle of *Origin* indicates just how important the notion of racial competition was in Darwin's scientific works: *The Preservation of Favoured Races in the Struggle for Life*. Here, he tells us precisely how important the struggle for existence between *races* is to his work.

In the second chapter of *Origin*, for instance, Darwin hammered home the notion that "dominance leads to greater dominance" when it came to animal and plant groups. Those groups already broad and diversified, able to dominate in different environments, will continue to vary and displace rival groups in more and more habitats. He repeated this group variation and competition theme in the 1868 first book of his evolutionary "trilogy," *Variation in Animals and Plants Under Domestication*. In the animal and plant kingdoms, on average and over time, the rich get richer.

Unsurprisingly, this theme of interracial difference emerges where Darwin did explicitly address humanity in *Descent of Man, and Selection in Relation to Sex*, published in 1871. In the center of Darwin's account, he leaned heavily on two painstaking studies. First, that of John Beddoe, physician and author of *Races of Britain*, published in 1862, the first "big data" project of the era, comparing traits such as stature, skin color, and eye color. And, second, that of Joseph Barnard Davis, author of *Crania Britannica*, published in 1865, who carefully compared cranial capacities, with Caucasians always coming out with the largest, "more perfect" crania, of course. While Darwin denied that quantitative comparisons between two individuals yielded much of worth, he upheld the beliefs expressed by both Beddoe and Davis that brain size and intellectual capability did correlate when it came to big groups like human races – "supported by the comparison of the skulls of savage and civilised races" with whites on the high end and dark-skinned Australians on the bottom (74–75). As well as variation and racial ranking, *Descent* attempts to account for racial extinction by appeals to these differences in racial intelligence, cultural flexibility, or "grade of civilization" (212). When the European colonizer contacts any Aboriginal population, wrote Darwin, "the struggle is short" and the Caucasian group always wins. To Darwin this seemed a natural law, another consequence of natural selection.

Anti-cruelty, for sure. Anti-slavery also. But there is little indication that Darwin held "all men are created equal" Enlightenment values. Colonialism may be a vast white scythe sweeping Aboriginal peoples off the globe, but Darwin seemed not to find this more than a bit troubling – the way a nineteenth-century geologist might find volcanic eruptions destroying island populations a bit troubling, but also an ordinary part of the natural world. He was a wealthy man near the social peak of an empire subjugating many across the globe, and he expressed very little interest in seeing that state of affairs change.

Still, it's a long way from these sentiments scribbled by a quiet man in his Victorian study to those expressed to Vernon Kellogg by German officers during the Great War. More frightening versions of Darwinism would take several other inputs that had little to do with Darwin. And the Third Reich's "Final Solution" drew not only from centuries of antisemitism – horrific pogroms of European Jewish communities pre-date Darwinism by more than half a millennium – but also from French, German, and American scholars, physicians, and politicians who ignored or disputed Darwin.

"Let the one with understanding solve the meaning of the number of the Beast" (Rev. 13.18; NLT)

To find the scientific inputs to the twentieth-century death camps, we have to rewind the story all the way back to the debates between Lord Monboddo and Lord Kames in the 1700s that I mentioned in Chapter 1. Monboddo argued that all human races converge into a single set of ancestors – monogenism – which may have been pre-human primates. Kames, like so many in the late 1700s, saw this version of monogenism as blasphemous. But instead of holding to an "Adam and Eve" beginning for humanity, Kames proposed polygenism – each human race had its own origin story, its own "Adam and Eve." Ultimately, this meant that, despite obvious similarities, what we commonly call *"the* human race" is several different breeds or even species. By the late 1700s, German and French "anthropologists," including the important philosopher Immanuel Kant, decided that humans should be divided into four or five – or maybe seven or nine or more, who can be sure? – distinct stems, species, sub-species, breeds, stocks, stirps, types, or races. No matter the uncertainty about the exact number, in every polygenic schema, whites always claim the privileged position, darker-skinned peoples always remain relegated to an inferior status (Figure 7.1). In a backhanded way, it granted the global slave trade with the accompanying destruction of Aboriginal populations – in other words, the politico-economic system making men like Kames and most European Empires so incredibly wealthy – a scientific blessing. This was just the natural order.

Mixing that order up was the problem.

French ethnographer Julian-Joseph Virey agreed, in a book read by Thomas Jefferson, among others. Once upon a time, Virey insisted in *The Natural History of Humanity*, published in 1801, original human stems *("genre")* featured their own distinct cultures, spoke their own distinct "mother tongues." But over time, these purities became "altered, corrupted by idioms that the mixture of various races" produced. Today, Virey mourned, one could literally hear the degeneration as pure languages belonging to each of the original pure human races mixed. This racial admixture *("les mélanges de races diverses")* signified society-wide corruption.

PROGRESSIVE DEVELOPMENT OF MAN.—(2) Evolution Illustrated with the Six Corresponding Living Forms.

Figure 7.1   Human races demonstrating "evolution" according to late-nineteenth-century interpretations of Darwinism.

Fear, both popular and scientific, that mélange/admixture led to degeneration increased over the nineteenth century. In the young United States, physician-scientists such as Samuel G. Morton and Josiah C. Nott (*Types of Mankind*, quoted above, was his idea) attempted to demonstrate White superiority and Black inferiority through large collections of skulls. If no one stepped in to halt intermarriage between races, Nott warned, the process of "hybridization" would lead to massive degeneration in the war between races.

Their contemporary, disgruntled French diplomat and novelist Joseph Arthur de Gobineau, saw evidence of racial admixture already degenerating Europe in his own day. Gobineau claimed he was driven to publish his *Essay on the Inequality of the Human Races* in 1855 because the death of nations occurred due to such degeneration. The political upheavals of 1848 signaled to him that his fellow French countrymen and women already started the slide, begun while interbreeding with enslaved workers on their plantations (secretly, he worried that he and his wife might pump the blood of enslaved French Caribbeans through their own veins). Only the blonde "German" or "Aryan" would rise above degradation. Everyone else? Well: "those who are not German are created to serve," he said. Gobineau's argument impressed Nott so much that he sponsored Henry Hotz, a Swiss immigrant Nott had met in Mobile, Alabama, to translate the work into English. Hotz agreed since he saw the presence of four major race stems interbreeding in the USA as a major threat. The Hotz–Nott 1856 translation of Gobineau thrust the dangers of racial admixture into the limelight in the Anglo-American world and beyond for decades.

Sentiments such as these would find purchase in German-speaking Central Europe and sprout ever more fervently as the nineteenth century wore on. By its end, "Gobinism," promoting racial purity and Aryan supremacy, persisted in several Gobineau Circles (*Gobineau-Vereinigung*) blooming throughout the German intelligentsia. "Flight of the Valkyries" composer Richard Wagner became a fan. So did Friederich "Will to Power" Nietzsche.

Darwin could have nothing to do with these satanic ideas, which slithered around before *Origin of Species* in 1859. He had even less to do with a parallel fear of degeneration that leapt up in French psychiatric circles mid-century.

While agreeing with Gobineau's overall concept of degeneration, the Aryan race science bit bothered Dr. Bénédict-Augustin Morel, a psychiatrist as much interested in the anthropological as the psychological. Morel noted increasing amounts of mental illness in his young patients at Asile d'Aliénés de Maréville, an asylum in Nancy, France. (One patient in 1852 complained that he became a bloodthirsty wolf at night and begged to be taken outside and shot; Morel, hopefully for the better, did not oblige him.) And in Morel's next role directing the Asile d'Aliénés de Saint-Yon in Rouen, Normandy, Morel recognized an unmistakable trend. Heads were changing; brains, shrinking. It was disturbing.

But he insisted it had nothing to do with racial admixture. It was the fault of the modern city. With the growth of industry, urban environments squeezed the populace both metaphorically and literally, Morel and his colleagues conjectured. They saw the first disturbing hints of a rapidly changing environment in a wave of degenerate children. Nevertheless, interracial fertility still proved that people were people, regardless of race. Mixed-race folks would show no biological drop-off, no diminishment from the level of their parents, against what Gobineau, and so many like him, assumed. Limited fertility and sterility in those he called "degenerate varieties" occurred because of the stresses of modern life.

He wondered how widespread this phenomenon was. So, during the 1850s and 60s, Morel traveled throughout France, visiting asylums. Then beyond. He assessed patients and conferred with physicians. Eventually he became convinced that therapy, no matter what kind, wasn't going to work; he was not seeing something curable. Those locked up in asylums, Morel stated, regretfully, were victims of "hereditary predispositions" visible in the head. No matter the proximate causes of mental illness, there was something underneath it that was inherited, something physical, something hard, unalterable.

It's true that Morel initially pointed to environmental conditions, rapid urbanization for one, as the source of degeneration. But over time, Morel changed his emphasis. He clearly found a degenerated *state*, a particular variety of humanity, something like a lower valley below the plateau of normalcy. Environmental stress proved to be only the trigger, the push over the edge. Normal people, Morel thought, would not suffer the way that abnormal people, the people with hereditary dispositions, did. And Morel really did mean "normal." He couched his own study alongside the "normal man" concept championed by his contemporary, famed Belgian mathematician Adolphe Quêtelet.

In his own breakthrough book, published in 1835, Quêtelet stressed that alongside normal or "average" individuals there existed a substantial number who did not qualify as normal. Unfortunately, it is from this number that most crime originates. Quêtelet claimed that "moral illnesses" can be contagious. Perhaps they're hereditary. In any case, if the medical system didn't control these abnormalities, cities would fill with criminals before long.

Indeed, Morel confirmed Quêtelet's suspicions. By the mid-nineteenth century, crime rose in parallel to the growing number of patients in asylum populations. Government ministers were alarmed, physicians were alarmed, police were alarmed, even religious figures were alarmed. Society showed every sign of decay, degeneration.

So, though it's true that Morel disavowed Gobineau's racism and emphasized the commonality of human races (monogenism), he and Quêtelet merely transformed a degeneration concept already freighted with plenty of racism into something scientific, connected with precise data, with heads and skulls and statistics. This is the point that lasted. Degeneration of civilization by allowing these individuals to roam free, commit crimes, and reproduce, increasing their future numbers, would inevitably follow.

To be clear, this isn't exactly Social *Darwinism*. Social Darwinism rested on an assumption that the status quo represented an expression of natural, biological order. Inequities in class, race, status, political power, economics, and so on merely indicated that the best people were rising to the top. Mid-twentieth-century US historian Richard Hofstadter once commented that American corporate capitalism in the post-Civil War period represented a caricature of the Darwinian struggle for existence. That's why we call it Social Darwinism.

But the fear expressed by psychologists like Morel turned out to be the inverse. From the perspective of British and French observers, the *worst* sorts of people (e.g., the poor, the Irish, the Eastern European, the degenerated) were winning the long-term struggle for existence. Darwinism – real biological Darwinism – had little to do with wealth or status, but with numbers of offspring. The impoverished, the undesirable, and the hereditarily tainted, according to the fears of physicians and scientists in the last half of the nineteenth century, left behind more babies than the healthy and wealthy. As one anxious commentator put it, this wasn't the extension of Darwinism into the social, but evidence of the *failure* of natural selection to operate among humans. If the predictions of Quêtelet ran true, that also meant that the criminally defective would increase as well.

The Architects of Ruin

One obvious solution might be to address poverty, illness, and crime through more health care workers, better funded educational systems, safe food and drinking water, job opportunities, and so on. Societies, in other words, that provided for "the least of these." Those who feared degeneration (e.g., Virey, Gobineau, Nott) insisted those interventions could never work. Defectives were hereditarily tainted and passed their unfitness on as discrete traits. The problem wasn't the external social environment. It was something carried inside.

Few pushed this view harder than Texas frontier physician, and correspondent with Darwin, Dr. Gideon Lincecum. He offered a straightforward solution as well, one which he lobbied heavily for and attempted multiple times to codify into Texas law through the mid-1800s. Cut out the defective hereditary material. Emasculate the offenders. The surgeon's knife would chop down crime, slice illness, sever all aspects of tainted hereditary once and for all. To save society, righteous physicians are going to have to dispense surgical justice on some people's genitals.

Francis Galton, cousin to Darwin, thoroughly agreed that good and bad traits passed on – he was much firmer than his cousin Charles that the environment couldn't alter these fundamental units of heredity (Figure 7.2). And he feared that the struggle for existence was no longer severe enough with humans to weed out degenerates. He began promoting his views in popular outlets in the 1860s and 70s. And he found strong support among other prominent government figures. Political theorist and Liberal Party statesman William Rathbone Greg, for instance. (We last met Greg at Plinian Society meetings in Edinburgh alongside Darwin long ago.) Greg decried degeneration as identified by Morel visible amongst both Britain's impoverished masses and its nobility. In a perfect world, Greg wrote in 1868, the "sick, the tainted, and the maimed" simply would refuse to procreate. Or, if they wouldn't abstain willingly, the British government would bar them from making more of their kind. Though in all other respects a proponent of fewer laws and freer trade, when it came to allowing the degenerate to breed, Greg advocated costly government interventions. Examinations, for instance. Those who didn't pass the examinations because of "damaged or inferior temperament" might be "eliminated."

Galton may have agreed with Greg. But he did not support the American idea of emasculation or asexualization. In 1883, he coined the term "eugenics" to describe a more humane path toward hereditary health. Promote larger families produced by the best sorts of people. People like him, the Darwins, and

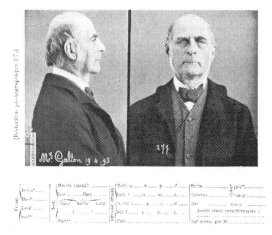

Figure 7.2   Sir Francis Galton as the biological repository of multigenerational accomplishment.

their Wedgwood cousins. The industrious, leisured elite, in other words. (Charles didn't let Galton down with ten children, seven of whom lived to adulthood. Paradoxically, Galton remained childless.)

Though he pushed and pushed for his eugenics policies to be taken seriously, it took until 1907 to formalize the Eugenics Educational Society in London, with the now elderly Galton serving as president and 22-year-old Sybil Katherine (née Burney) Gotto (after 1917, Neville-Rolfe by her second marriage) the secretary, organizer, and boundless source of reformist energy. Neville-Rolfe kept the British eugenics movement closely tied to her Committee of the Moral Education League, her National Council for Combating Venereal Disease, and her National Council for the Unmarried Mother and her Child. These were social organizations pushing for clean living, middle-class values, and investment in moral education, not draconian avenues for social control by surgeons' scalpels. In 1913, she successfully convinced Parliament to consider the Mental Deficiency Act, the culmination of over three decades of agitation by Galton, which would have criminalized marriages between those deemed unfit.

Galton might seem like an obvious tie between Darwin and twentieth-century developments leading to death camps. Numerous books and articles on eugenics will hint at this link. Except history didn't turn out that way. Neville-Rolfe's version prevailed; the more controlling medical aspects of eugenics never came to pass in Britain. No real marriage restrictions. No increase in castrations. Most direct links from Darwin's work to the Darwinism championed by, for instance, the German High Command in World War I as recorded by Kellogg are severed right here at this historical fact. Truth be told, "Darwinism" supposedly motivating the atrocities of the Third Reich is even further removed.

How, then, did those terrible things come to pass with even the thinnest veneer of "survival of the fittest" science still clinging? Put another way, why do people believe Darwin's ideas have anything to do with death camps?

Instead of Darwin or Galton, we must return to Darwin's contemporary, Texas physician Gideon Lincecum, for an answer; it's his initiative that came to pass. That's right: the road to Auschwitz ran not through Darwin's Down House in Britain, but through various workaday American physicians, asylum directors, and prison wardens who, without legal clearance, nonetheless segregated, sterilized, castrated, and castigated incarcerated men through the 1890s.

Lincecum's law never passed, but for decades in medical journals and legal conferences, American physicians mimicked him, preaching asexualization for criminals, alcoholics, and anyone showing the marks of degeneration. And some powerful American politicians seemed all too eager to put some of these concepts into practice. By 1897, Connecticut made it illegal for those bearing hereditary taint to marry, and Michigan unsuccessfully attempted to institute a full-bore sterilization law. Also that year, Dr. Albert J. Ochsner, the well-respected Chicago surgeon, recounted his use of a new European surgery – the vasectomy – and recommended using it to halt the production of habitual criminals.

Within months, Harry Sharp, the appropriately named physician at the Indiana Reformatory in Clarksville, Indiana, just across the Ohio River from Louisville, Kentucky, began employing precisely this surgery on inmates he deemed morally defective. By 1907, when the legislature of the state of Indiana, pressured by the governor, finally passed the world's first eugenics law allowing these surgeries to take place, Sharp had already sterilized 176. He claimed they gave consent.

State after state from coast to coast passed eugenic legislation. American physicians sterilized at least 30,000 in the decades between the First and Second World Wars. They would sterilize tens of thousands more after them. They focused on the mentally handicapped, mainly. And it's true, the big American eugenics meetings featured statues of Galton and Darwin, claiming them both as forefathers. In reality, however, the American eugenicists were more influenced by the advocacy of contemporary bigots like Madison Grant, whose *The Passing of the Great Race, or, the Racial Basis of European History*, published in 1916, bemoaned degeneration in whiteness itself. While "Nordics" – Gobineau's old Aryans – contributed to society, "Alpines" and "Mediterraneans" largely sapped whatever vital strength lay in the essence of Caucasianness. Stopping the degeneration of society would mean stopping breeding among these white undesirables as well. Grant's lament regarding the master race ricocheted in the post-WW1 era through a rising Ku Klux Klan in the Midwest to the Race Betterment Society in California. Darwin proved anathema to these pseudo-Christian groups.

Harvard University-trained geneticist Charles Davenport anchored a less overtly racist, more scientific-seeming end of an official American eugenics movement from Cold Spring Harbor, New York, on Long Island, not far outside of New York City. He and a large team of researchers ran the Eugenics Records Office, tabulating the unfit and pressing for more draconian laws preventing their reproduction. Davenport successfully stitched together a whole national and international eugenics movement that targeted the hereditary taint they insisted led to crime, mental illness, behaviors like alcoholism and prostitution, poor performance in school, and interracial marriage.

But it was Davenport's chief henchman, Harry H. Laughlin, who proved even more important for the story that we now realize ended in severe immigration restrictions in the USA and gas chambers in Europe. It was Laughlin who wrote the "Model Sterilization Law" in the United States.

That law interested Albert Johnson, Washington state Republican, who made his run for Congress on a platform of pan-Aryanism in 1913. With advice, testimony, and encouragement from Laughlin, Johnson cobbled together the

1924 Immigration Act, better known as the Johnson–Reed Act, that throttled the entry of non-white "aliens" into the United States. As Johnson portrayed it, non-whites were far more likely to bring hereditary disease to the country, more likely to become "takers," more likely to produce crime. The act stood largely unchallenged for five presidential administrations over four decades. (More recent politicians, like Trump's first Secretary of State Jefferson B. Sessions III of Alabama, reflected wistfully on the Johnson–Reed Act and regretted its 1965 repeal.)

Laughlin also became the main go-between crafting legislation in Germany that, in the hands of the Third Reich, structured a eugenics program that went well beyond what even American society tolerated. The University of Heidelberg awarded Laughlin an honorary MD in 1936 for his sterilization policy work. Dean of the Heidelberg Faculty of Medicine, Dr. Carl Schneider, lauded Laughlin for guiding the National Socialist Party (Nazis) in formulating their own sterilization laws. Laughlin wrote overjoyed that the Nazi physicians found common cause with Americans in the quest to clean the white race.

The *Aktion T-4* program, launched in the wake of this exchange, sterilized, then executed, tens of thousands of men and women deemed hereditary degenerates before being formally discontinued in 1941 (Figure 7.3). Informally, however, the program continued. German and Austrian physicians and other healthcare workers under the direction of Dr. Viktor Hermann Brack, and with Hitler's personal physician Dr. Karl Brandt clearing the way politically, systematically murdered perhaps 300,000 people by 1945. In six such eugenics asylums – Brandenburg, Grafeneck, Bernburg, Sonnenstein, Hartheim, and Hadamar – Brack's underlings gassed thousands of those deemed unfit, in makeshift showers. Brack soon thereafter collaborated with *SS-Obersturmbannführer* Otto Adolf Eichmann to industrialize this notorious process across Central Europe in service of the "Final Solution."

This litany of complicit figures stands woefully incomplete, of course. Thousands of ordinary men and women instituted the policies and conducted the procedures that resulted in discrimination, castration, vasectomy, tubal ligation, hysterectomy, immigration restriction, and eventually euthanasia. The vast majority of them based their work not on anything written by Charles Darwin.

Figure 7.3   Third Reich eugenics propaganda "How it ends" showing the displacement of "superior" Germans by the "inferior" in just a few decades unless drastic measures were taken to alter the population of degenerates.

That doesn't mean Darwin is off the hook, exactly. But if we insist on placing Darwin on a line that led in the twentieth century to some of the most thoroughly organized and most heinous crimes committed against the human species, then we must acknowledge that, even in the problematic letters and passages I spelled out above, Darwin reflected actions *already* part-and-parcel of Anglo-American and European colonization, deep-seated racism and xenophobia, and imperial conflict. It's a very, very thick line. It started centuries before Darwin. That line bleeds all over us even today.

Despite what Vernon Kellogg recorded in Belgium in 1916 and what William Jennings Bryan feared would come of American students learning Darwinism a decade later, taking aim at the use of the phrases like the "struggle for existence" and "survival of the fittest" in the *Origin of Species* remains a sloppy post-hoc justification for genocide. No phrase used by a quiet naturalist fiddling with orchids, pigeons, and earthworms in rural England was needed to unchain humanity's demons. If anything, we can imagine Darwin sadly shaking his head. Perhaps this is what he warned militant atheists Edward Aveling and Ludwig Büchner about in September 1881. The struggle for existence does not remove the moral imperative: "man can do his duty," insisted Darwin.

Or, as "bulldog" T. H. Huxley phrased it in his 1897 "Evolution and ethics":

> Let us understand, once for all, that the ethical progress of society depends, not on imitating the cosmic process [i.e., evolution by selection], still less in running away from it, but in *combating* it. . . .
>
> We are grown men, and must play the man . . . cherishing the good that falls in our way, and bearing the evil, in and around us, with stout hearts set on diminishing it. So far, we all may strive in one faith toward one hope . . . (83, 86)

He surely couldn't feel the prick of historical irony, but Huxley, ardent follower of Darwin that he was, preached this *fin-de-siècle* jeremiad against unrestrained scientific hubris baptized in the poison of colonialism and imperialism at the very moment American physicians stood, scalpels in hand, on the edge of cutting.

Concluding Remarks

The Legend Machine

Historian Everett Mendelsohn was intrigued. In the middle of writing a review of an annual survey of academic publications in the History of Science, he marveled that an article in that volume contained almost 40 pages' worth of references to works on Darwin published in just the years between 1959 and 1963. Almost 200 works published in a handful of years – no single figure in the history of science commanded such an impressive academic following. Yet Mendelsohn noted that, paradoxically, no one had written a proper biography of Darwin by 1965. Oh sure, there was *commentary*. Lots of commentary. But so many of the authors were retired biologists who had a tendency toward hagiography or, the opposite, with axes to grind.

In the aftermath of the Second World War, descendants of Charles and Emma Darwin donated manuscripts, correspondence, and Darwin ephemera to Cambridge University Library, which ever after would serve as the de facto hub of Darwin studies. Nora Barlow, one of those descendants, assembled Darwin's original autobiography, reinserting passages that son Francis Darwin purged decades earlier to avoid offending friends, family, and colleagues after his father's death. But few historians or philosophers had spelunked what proved to be a cavernous mine of archival material by the mid-twentieth century.

It hardly mattered. The books and articles just kept flowing. From Mendelsohn's vantage point, the biologist-dominated celebrations around the centenary of *On the Origin of Species* and Darwin's 150th birthday in

1959 were the precipitating events. There were so many of them. And they clearly uncorked a veritable deluge of Darwiniana that, with no signs of slowing by the mid-1960s, indicated an entire "Darwin industry" had appeared.

The term stuck.

A decade later in 1974, philosopher Michael Ruse subtly suggested that scholars were cashing in on the Great Man Theory of the History of Biology. He offered a critical assessment of the Darwin industry (it turned out only to be its first generation), given the uneven quality of publications. Ever after, "Darwin industry" carried that connotation: less-than-innovative scholarship created to feed what sure appeared to be a brewing culture war. Some of it was only loosely connected to Darwin even if it carried the "Darwinism" label. The British naturalist just happened to lie at the symbolic core of the dust-up.

More so than almost anyone, formidable ornithologist Ernst Mayr of the American Museum of Natural History and, later, director of Harvard's Museum of Comparative Zoology rebranded mid-twentieth-century evolutionary biology as the direct descendent of Darwin's work a century earlier (Figure C.1). Paradoxically, Mayr massaged the degree of modification in that

Figure C.1   Ernst Mayr (1904–2005) did more than most anyone to cement Charles Darwin's legacy as a peerless scientific progenitor of modern evolutionary theory.

intellectual tree of descent at every turn. He downplayed the opposition between Mendelism and Darwinism from earlier in the century and deflected references to Lamarckian mechanisms in Darwin's work – despite that Darwin tended *more* toward the soft, environmentally shaped inheritance of Lamarck after *Variation in Animals and Plants Under Domestication*. To this end, Mayr spearheaded reissuing a facsimile of the first edition of the *Origin* as "true" Darwinism. He ventured into Anglo-American History and Philosophy of Biology (which he contrasted with Theoretical Biology dominated by Europeans with less fealty to Darwinism) and sharpened the profundity of Darwin's contributions toward biological method as well as theory. Mayr also drew on his core American network of George G. Simpson (paleontology), Theodosius Dobzhansky (population genetics), and G. Ledyard Stebbins (botany; collectively dubbed the "Four Horsemen"), as well as outspoken British evolutionists Julian Huxley (grandson of T. H. Huxley), E. B. Ford, C. D. Darlington, and J. B. S. Haldane (all polymathic geneticists) to market the "Modern" or "neo-Darwinian" Synthesis. A 1970s conference and resultant book solidified the historical narrative: from Down House, lacking a university position or a real laboratory, without the need to write grants or publish in peer-reviewed journals, saintly, under-appreciated and misunderstood Charles Darwin poured the foundations for today's university-based, laboratory-housed, grant-driven, and postgraduate-student-dependent biology, paleontology, even anthropology. To cap it off, as if ordered from the Great Beyond by Darwin the secular saint, in 1979 the BBC produced a high-quality dramatization of Darwin's *Beagle* voyage and his evolutionary discoveries. They ended up codifying Mayr's vision on the TV screen. True, there was some professional nuance here, but by mid-century *evolution* just was Darwin's 1859 insight plus genetics. (And neither Darwin's "pangenes" nor the "beanbag" version of genetics, either.)

Ruse's 1974 review proved prescient. Before the decade was out, a glut of evolution-adjacent books erupted fueling culture wars, including British zoologist Richard Dawkins' *The Selfish Gene* published in 1976, Alabama entomologist E. O. Wilson's *Sociobiology: The New Synthesis* published in 1975, and mathematician Douglass Hofstadter's *Gödel, Escher, Bach* published in 1979, among others. Collectively, they painted a picture wherein human aesthetics, religion, history, even psychology was illusory, where mindless genes danced us and our values around like marionettes on strings – which was how literally *Time Magazine* interpreted this body of literature in their 1 August 1977 cover story. They weren't stretching the metaphor by much, given quotes like the one Harvard biologist Robert Trivers offered: "Sooner or later, political science, law, economics, psychology, psychiatry and anthropology will all be branches of [Wilson's] sociobiology." What these had to do with *Darwin*, few could say. But, by the 1970s, the Darwin industry seemed to have unspooled its tendrils beyond depictions of the nineteenth-century British gentleman naturalist. It had more to do with, on the one hand, growing imperialism by the well-funded bio-medical sciences and, on the other, flag-waving in protest of political and religious conservatives.

Given how loud, aggressive, and politically organized those religious conservatives were becoming, perhaps the 1970s Darwin industry partisans could be forgiven. "Scientific" Creationism swam onto the scene in the United States for the first time since the *Scopes* trial in 1925 in the murky waters of the Reagan and Thatcher regimes. Creationists threatened to bend the powers of the state in service of teaching religion alongside or instead of evolution in the science classroom. In the 1981 *McLean v. Arkansas* case, a broad spectrum of scholars successfully pushed back, but the Creationists had made significant inroads in state governments and educational boards.

Coincidentally, the trial overlapped with the 1982 centennial of Darwin's death. Not surprisingly, academic works on Darwin and evolution erupted in the tumult over the trial and the Darwin celebration, driving a literature wave larger than the one in 1959–63. This time, a cadre of humanists in many disciplines joined the scientists. Among other places, they drew from Cambridge's ever-growing Darwin archive. The project at the crest of that wave, the over 1,100-page *The Darwinian Heritage* published in 1985, featured 32 top Darwin scholars who rehashed debates, filled in gaps, compiled massive bibliographies, and constructed the most complete narrative to date of evolution, Darwin, and Darwinism from the last two centuries.

Ironically, some authors of the 1980s Darwin industry began to undercut Mayr's narrative from Darwin to Modern neo-Darwinism. Historian Peter Bowler clarified that the so-called Darwinian Revolution did not occur in the nineteenth century and had little to do with Darwin's own works – it was literally a non-Darwinian revolution, an *eclipse* of Darwin's work, not its culmination. Dov Ospovat showed a great deal more continuity between the Natural Theology tradition and Darwin's Natural Selection, which supposedly overturned, destroyed, or replaced it. Older leftist critiques that the *Origin of Species* tracked too closely to oppressive socioeconomic trends of Victorian capitalism resurfaced. *The Descent of Man* promoted sexism. *The Expression of the Emotions* dehumanized the mentally disabled. Scholars recast Emma Darwin as the coddling anti-feminist, serving her conveniently always-ill gentleman scientist. And, inevitably, by the late-1980s, it became clear that the eugenics movement that sterilized tens of thousands in the United States and planted the seeds of Holocaust in Germany originated in Francis Galton's reading of "survival of the fittest" in his cousin Darwin's work. Other historians like Bill Provine and Gar Allen unpacked how fierce was the anti-Darwinism of the early genetics movement and how difficult it was to see Darwin's influence on any developments in modern biology prior to Ernst Mayr's reframing – which just so happened to coincide with the hyped 1959 centenary celebrations that publicly featured the Four Horsemen and their British allies. Perhaps most tellingly, two of the most accomplished Darwin scholars in the world, Adrian Desmond and James Moore, collaborated to produce in 1991 the greatest biography of Darwin yet written. But they subtitled it *The Life of a Tormented Evolutionist*. The Darwin industry veered away from honorifics of earlier generations and moved, on the one hand, deeper into minutia about Darwin's work and life and, on the other, away from centering Darwin as the master figure in evolutionary theory.

As the twentieth century ended, the Darwin industry appeared to lose some of its earlier steam. Evolution and Christianity looked somewhat less at-odds than at any point in the previous century. Scientific Creationism had been discredited. With the collapse of the Soviet Union, "godless" evolutionary science no longer loomed as part of an anti-Western plot. In a formal statement to the Pontifical Academy of Sciences in October 1996, Pope John Paul II conceded that human bodies (though not minds – Wallace still held ground here) shared common origins with non-human bodies. And frightening religious figures from the Near East also denounced "Western" ideas like evolution; suddenly, atheism and evolution appeared the lesser threat.

Proper historians and philosophers of the life sciences numbered in the hundreds by then and, rather than treading the same well-worn path of their exclusively natural-science-trained predecessors, fanned out in their research beyond just the Down House-bound Sage. The "Science Wars," including mathematician Alan Sokal's mid-1990s hoax of left-winged postmodern cultural studies, may have led to a re-evaluation of projects challenging Darwin's status inside the academy as well. Janet Browne's two-volume biography of Darwin, a monumental and less controversial take than the Moore and Desmond volume – if anything, even further down into the Darwin weeds, given the degree to which Browne combed through the hard work of scholars on the Cambridge Darwin Correspondence Project – put a virtual endcap on the second generation of the Darwin industry.

But several events in the early twenty-first century stoked the industrial furnaces again. First, the completion of the Human Genome Project drew more media attention, grant money, and graduate students into the field. Second, a resurgent Creationism, now in the guise of Intelligent Design issuing from Seattle's well-funded Discovery Institute, began beating the war drums again. The resulting 2005 *Kitzmiller v. Dover Area School District* case became the most prominent religion versus science trial in a quarter century. Combined with a third factor, the George W. Bush administration's restrictions on promising stem cell research in summer 2001, it appeared the culture wars were dragging evolutionary biology back into the fray. In reaction, a confrontational New Atheism, anchored by some of the same people who had been populating the Darwin industry for four decades, erupted on blogs, social media, talk shows, podcasts, and in public discussions, lectures, and debates.

By coincidence, these developments roughly overlapped with yet another Darwin centenary in 2009, at the very same moment the financial crises of the late "oughties" spelled funding cuts for science and, especially, those humanists who study science. Publications on Charles Darwin, couched in a much more robust historical, sociological, economic, and philosophical context than ever before, fired out from a new and improved Darwin industry. Perhaps not surprisingly, Browne the master Darwin biographer tabulated over 1,000 books just with Darwin's name *in the title* published globally since 1985.

Yet she also sees a very different industry. Darwin is no longer central. No one denigrates the concepts of common descent, natural selection, and sexual selection. Far from it. Over a century and a half have corroborated Darwin's hypotheses over and over again from many different angles in geology, genetics, anthropology, and embryology. Even some parts of Darwin's theories that other biologists de-emphasized, even mocked at the turn of the twentieth century, turn out to be based on what we now know was solid evidence: that environment not only selects traits but likely impacts inherited traits (for this, we look beyond genetics to epigenetics).

This generation of the Darwin industry decentralizes Darwin because historians have grasped over the past decades that to understand Darwin, we must grasp how idiosyncratic was his path. (Even the book in front of you implies this throughout.) Without diminishing his accomplishments, we no longer lionize him as a genius, but stress the degree to which his contributions to human knowledge may have had more to do with relationships of status and sheer accidents of history than the weight of their content. In the case of Darwin, it was as much *who* he knew, the status of those individuals, and the degree to which that network would promote his ideas into an even larger scientific network already working for several generations with the concepts (common descent with modification, the Malthusian competition for resources, etc.) that launched Darwin into the pantheon occupied by single-named intellects: Newton, Aristotle, Curie, Einstein, Galileo, and so on.

As the greatest champions and, ironically, critics of the Darwin industry remind us – Michael Ruse, Mary Midgely, and John C. Greene, for instance – to reach that pantheon, it's not enough to have good scientific ideas. (Alfred Russel Wallace had good scientific ideas, and we have to use all three of his names to spark recognition even among those who have heard of him.) Icons whose names ring down the ages symbolize something greater than themselves, greater than any scientific concept.

This brings us back around to the "earthworm stone" I referenced at the beginning of the book, still poking out of the ground at Down House.

The ideas of J. Robert Oppenheimer, Melitta Bentz, Arthur Holly Compton, Grace Hopper, and Tim Berners-Lee, to grab just a handful at random, arguably have a greater impact on the course of geo-politics and our daily lives than anything proposed by Darwin. No one preserved Grace Hopper's home. No one makes a pilgrimage to Melitta Bentz's grave. No one celebrates Oppenheimer's birthdays around the globe every 50 years. No one assigns Berners-Lee's key texts in a classroom. And there aren't courses in, say, Compton's impact on literature or pop culture. Yet many of us do these things in honor of Charles Darwin because the icon Charles Darwin has come to symbolize, either now or in the recent past, a whole litany of sociopolitical ideas and practices: the benefits of brutal competition in business, politics, and athletics, the inevitability of technological progress even to the elimination of humanity itself, the victory of European colonizers over "unfit" races around the world, the relative importance of engineering and computerized technology over the traditional liberal arts, a justification for eliminating the mentally and physically disabled, and the foolishness of evangelical Christianity and the people who devote themselves to it. It doesn't matter that Darwin didn't say or stand for these things. It doesn't even matter that present-day neo-Darwinism as defended by life scientists, historians, and philosophers all over the globe has nothing to do with these things. Darwin and Darwinism transcend their given meanings. The ideas have grown beyond their original containers, out of the pot into which Darwin's "I think..." tree was planted.

Darwin, in other words, transmutated into a religious icon. That earthworm stone behind Down House, then, is a kind of relic.

Witness, for instance, the vitriol aimed at American anthropologist Agustín Fuentes in 2021 at the 150th anniversary of the first edition of *The Descent of Man* (curiously, *not* the edition we read today). Fuentes excoriated the book in *Science*, calling it "problematic, prejudiced, and injurious," a text that "offer[ed] justification of empire and colonialism, and genocide...." Darwin's racism and sexism, said Fuentes, blinded him to both "data" and his positive interactions with intelligent women and people of African and South American descent. Aghast, a multidisciplinary collective of British, European, and American scholars quickly assembled to lash out at Fuentes (a sign all by itself – hardly any idea spurs academics across disciplinary boundaries to assemble quickly). Far from being a bog-standard Victorian racist, they claimed, Darwin maimed slavery-justifying polygenism, inspiring anthropologists of later generations, even the vaunted anti-racist anthropologist Franz Boas (this isn't true, by the way). Fuentes' insinuation to the contrary opened the proverbial microphones to "anti-evolution voices" and dissuaded the "more gender and ethnically diverse" evolutionary scientists of the future from pursuing their scientific passions. Fuentes may have had facts on his side (see Chapter 7), and it's hard to see a *critique* of Darwin's racism and sexism turning away today's students from the life sciences. But that really wasn't the point. The point was Fuentes had slandered the icon.

As much as we understand about Charles Darwin, he more than almost anyone else would hate the glare of this limelight.

Or perhaps he wouldn't. Because the misunderstandings that accrete around Darwin and Darwinism – a handful of which I addressed in this book – mean that at least people still engage with his ideas. They still care. The young man who popped beetles in his mouth because he didn't want to drop a specimen he could brag about later, who never got comfortable with the sea despite the nearly five-year circumnavigation, who could shoot a bird but hated blood and opposed vivisection, who never made MD at Edinburgh nor sat the Tripos at Cambridge, who got emotional seeing the casual horrors of slavery that his fellow scientists saw as *de rigeur*, who rarely made it through a month of his life without experiencing debilitating illness, who couldn't overcome his anxiety to attend his own father's or daughter's funerals, who stared into the abyss of time and death and still said there was something beautiful to be found in a tangled bank or a pigeon feather or an insectivorous leaf, that he nevertheless wanted to *do his duty* even if all ends in dust with no higher purpose – that man would be perplexed and a little amused, probably, that here we are a century and a half later still digging through the half-legible scrawl of his letters and the winding circuity of his overstuffed manuscripts, writing books about understanding *him*.

Summary of Common Misunderstandings

Darwin discovered evolution on the Galápagos Islands. The concept of "evolution" long pre-dates Charles Darwin. French scholars, especially, elucidated common descent with modification for decades. Even his grandfather, Erasmus, wrote about evolution extensively in popular works of the eighteenth century. While Darwin and several other HMS *Beagle* crew members, including Captain FitzRoy, captured specimens on a handful of Galápagos islands, the birds gathered there and identified by ornithologist John Gould in London did not suggest "evolution" to Charles Darwin. When Alfred Russel Wallace read Darwin's account of his 35 days on the Galápagos Archipelago, Wallace found Darwin's explanation lacking and began to piece together his own evolutionary explanation using what he believed Darwin couldn't see there.

Alfred Russel Wallace independently arrived at the same evolutionary theory as Charles Darwin. By 1858, Wallace regarded descent with modification as a more plausible explanation for the diversity of animal and plant life on the planet than Special Creation events. He envisioned a principle of growth and divergence held back by the struggle for existence as outlined by Rev. Malthus a half-century earlier, just as Darwin would. Yet Wallace also stressed selection acting against whole groups as much as against individuals. This is one point on which Darwin and Wallace emphasized different facets of evolution. Wallace also was not convinced that domestication gave us any insight into the process of evolution, unlike Darwin, who anchored his *Origin of Species* on that analogy. By the end of Darwin's life, he and Wallace diverged on an even larger number of points.

Darwin's Big Idea was merely natural selection. Their other points of disagreement reveal the expansiveness of Darwin's theory beyond merely natural selection. Darwin indicated non-adaptive or neutral change, which Wallace rejected, and sexual selection or female mate choice, which Wallace also regarded with skepticism. Wallace also rejected the idea of including human mental processes under the governance of natural selection, unlike Darwin, who wrote *The Expression of the Emotions in Man and Animals* in 1872 to smooth over the apparent gaps between the mental and emotional lives of animals and humans. To shore up criticisms regarding variation and inheritance, Darwin also dredged up the century-old concept of pangenesis floated by Pierre Maupertuis that identified gemmules as particles altered by the environment that could be inherited with their acquired alterations in the next generation.

Darwin promoted atheism. Though he engaged with other figures regarding his religious beliefs, he refused to commit definitively to any one position – aside from "unknowing" or, in the term coined by T. H. Huxley, "agnosticism." Atheists confronted Darwin, asking him to clarify his beliefs. He rejected their assertions that he confess godlessness. Among other reasons he refused to embrace atheism, he cited the faith of millions of others, including his mother and wife, and the existence of the Universe at all. *Acting as if* belief held weight seemed to create a social cohesiveness that Darwin feared might be lost without it. Additionally, throughout his time in Downe, the Darwin family actively supported through time and money the local church and affiliated charity groups, evangelical temperance reformers, and encouraged a South American missionary society.

Darwin's theory led to the death camps of the twentieth century. Darwin's record on human race is not clean. And he seemed to be only mildly troubled that European empires were destroying Aboriginal populations across the globe. Yet he deeply rejected cruelty toward humans and animals, supported abolition of the slave trade, much like other members of his family, and hoped for the cessation of slavery more generally. That any endorsement of a "war of the races" followed from Darwin's adoption of Herbert Spencer's term "survival of the fittest" can be easily disproven. Works endorsing such a vision pre-dated Darwin's. Many of those older works supported polygenism – the idea that humans do not share common ancestors. Darwin staunchly supported monogenism, that all races of humans share common ancestry in mind and body. And most individuals arguing for the extermination of other races did so without ever endorsing or even referencing Darwin's ideas. The truth is the death camps had their origins in several places that have nothing to do with Darwin, including American practices of emasculating criminals and sterilizing the unwanted in the nineteenth century, and in camps such as the World War I–era German colonial project operated in Namibia, in southwest Africa, where soldiers murdered tens of thousands long before the Holocaust proper and without any "survival of the fittest" justification.

References

DCP reference numbers in the text are quotes from private letters sourced from the Darwin Correspondence Project at the University of Cambridge Library, www.darwinproject.ac.uk.

Chapter 1

Anderson, K. 1999. The weather prophets: science and reputation in Victorian meteorology. *History of Science* 37: 179–216.

Averby, K. 2021. A long-lost grand house with a fascinating past. *East London and West Essex, Guardian Series*, 8 May. www.guardian-series.co.uk/news/192855 91.grand-east-london-house-home-one-first-infant-schools/

Barlow, E. N. 1946 [1845]. *Charles Darwin and the Voyage of the* Beagle. New York: Philosophical Library.

Browne, J. and M. Neve. 1989. *Introduction to Charles Darwin's* Voyage of the HMS *Beagle*. London: Penguin Books.

Burnett, J. (Lord Monboddo). 1797. *Antient Metaphysics; Volume Fifth: Containing the History of Man in the Civilized State*. London: Bell & Bradfute & T. Cadell, Jr. & W. Davies.

Burstyn, H. L. 1975. If Darwin wasn't the *Beagle*'s naturalist, why was he on board? *British Journal for the History of Science* 8: 62–9.

Darwin, C. 2004 [1871/1879]. *The Descent of Man; or Selection in Relation to Sex*, 2nd ed. New York: Penguin Classics.

Darwin, C. 2009 [1872/1890]. *The Expression of the Emotions in Man and Animals*, 2nd ed. New York: Penguin Classics.

Darwin, C. R. 1929. *The Autobiography of Charles Darwin; with Two Appendices* (Edited by F. Darwin). London: Watts & Co.

Darwin, E. 1791 *The Botanic Garden: A Poem in Two Parts*. London: J. Johnson.

Darwin, E. 1809 [1794–96]. *Zoonomia: or the Laws of Organic Life*, Vol. 1, 3rd American ed. Boston, MA: Thomas & Andrews.

Darwin, E. 1804. *The Temple of Nature; or, The Origin of Society*. American ed. Baltimore, MD: Butler and Bonsal & Niles.

FitzRoy, R. 1839. *Narrative of the Surveying Voyages of His Majesty's Ships* Adventure *and* Beagle *Between the Years 1826 and 1836*. London: Henry Colburn.

Humboldt, A. von and Bonpland, A. 2011. *Personal Narrative of Travels to the Equinoctial Regions of the New Continent: During the Years 1799–1804*. Translated by H. M. Williams. Cambridge: Cambridge University Press.

Keynes, R. 2003. *Fossils, Finches, and Fuegians: Darwin's Adventures and Discoveries on the Beagle*. New York: Oxford University Press.

Lyell, C. 1830–1833. *Principles of Geology: Being an Attempt to Explain the Former Changes of the Earth's Surface, by Reference to Causes Now in Operation*. 3 vols. London: John Murray.

McNish, J. 2018. John Edmonstone: the man who taught Darwin taxidermy. Natural History Museum London. www.nhm.ac.uk/discover/john-edmonstone-the-man-who-taught-darwin-taxidermy.html

Paley, W. 1827. *The Works of William Paley, D. D., Archdeacon of Carlisle*. Edinburgh: Printed at the University Press, for P. Brown and T. & W. Nelson.

Sargent, J. S. 2017. Charles Waterton – South America, Guyana, Mibiri Creek. *Overtown Miscellany*. https://overtown.org.uk/cw/Charles_Waterton/demerara-3.htm

Stephens, J. F. 1829–32. *Illustrations of British Entomology*. London: Baldwin and Cradock.

van Wyhe, J. 2009. Charles Darwin's Cambridge life, 1828–1831. *Journal of Cambridge Studies* 4: 2–13.

Viens, R. 2013. Darwin's apprentice – Syms Covington. *The Beagle Project.* https://beagleproject.wordpress.com/2013/06/30/darwins-apprentice-syms-covington/

Wakefield Museums & Libraries. 2020. Charles Waterton and slavery. Wakefield Council. http://wakefieldmuseumsandlibraries.blogspot.com/2020/11/black-history-month-charles-waterton.html

Waterton, C. 1909 [1825]. *Wanderings in South America, the North-West of the United States, and the Antilles in the Years 1812, 1816, 1820, and 1824.* New York: Sturgis & Walton.

Williams, L. 2018. Edinburgh's part in the slave trade. Historic Environment Scotland. https://blog.historicenvironment.scot/2018/11/edinburghs-part-slave-trade/

Young, V. and E. Tenkate. 1998. Expanding worlds. *The Journal of Syms Covington*, Australian Science Archives Project. www.asap.unimelb.edu.au/bsparcs/covingto/chap_1.htm

Chapter 2

Adoum, J. E. 1982. The Galapagos Islands: the origin of '*The Origin*'. *The UNESCO Courier* 35. https://unesdoc.unesco.org/ark:/48223/pf0000048983.locale=en

Anon [Grant, R. or Jameson, R]. 1826. Observations on the nature and importance of geology. *Edinburgh New Philosophical Journal* 1: 293–302

Bowler, P. J. 2003. *Evolution: The History of an Idea.* Berkeley, CA: University of California Press.

Colnett, J. 1798. *A Voyage to the South Atlantic and Round Cape Horn into the Pacific Ocean.* London: Bennett.

Darwin, C. 1839. *Narrative of the Surveying Voyages of His Majesty's Ships* Adventure *and* Beagle *Between the Years 1826 and 1836 – Journal and Remarks, 1832–1836*, Vol. III. London: Henry Colburn.

Darwin, C. 1845. *Journal of Researches into the Natural History and Geology of the Countries Visited During the Voyage of H.M.S.* Beagle. 2nd ed. London: John Murray.

Darwin, C. 1859. *On the Origin of Species by Means of Natural Selection, or the Preservation of Favoured Races in the Struggle for Life.* London: Murray.

Darwin, C. 1958. *The Autobiography of Charles Darwin (1809–1882); with the Original Omissions Restored* (Edited and with appendix and notes by N. Barlow). London: Collins.

Descartes, R. 1853 [1637]. *Discourse on the Method of Rightly Conducting the Reason and Seeking Truth in the Sciences.* 2nd ed. (Translated by J. Veitch). London: Simpkin, Marshall, and Co.

de Maillet, B. 1750. *Telliamed: or, Discourses Between an Indian Philosopher and a French Missionary, on the Diminution of the Sea, the Formation of the Earth, the Origin of Men and Animals, and other Curious Subjects, Relating to Natural History and Philosophy* (English translation). London: Osborne.

FitzRoy, R. 1839. *Narrative of the Surveying Voyages of His Majesty's Ships* Adventure and Beagle *Between the Years 1826 and 1836.* London: Henry Colburn.

Geoffroy Saint-Hilaire E. 1830. *Principes de Philosophie Zoologique.* Paris: Pichon et Didier.

Gribbin, J. and M. Gribbin. 2003. *FitzRoy: The Remarkable Story of Darwin's Captain and the Invention of the Weather Forecast.* New Haven, CT: Yale University Press.

Lamarck, J. B. 1793. *Recherches sur les causes des principaux faits physiques.* [*Research into the Causes of Principal Physical Facts*]. Paris: Maradan

Lamarck, J. B. 1801–10. *Annuaires météorologiques.* [*Meteorologies*]. 11 vols. Paris: Deterville.

Lamarck, J. B. 1964 [1802]. *Hydrogéologie.* (Translated by A. V. Carozzi). Urbana, IL: University of Illinois Press.

Lamarck, J. B. 1809. *Philosophie Zoologique* [*Zoological Philosophy*]. Paris: Musée d'Histoire Naturelle.

Leclerc, G.-L., Comte de Buffon and B. Germain de Lacépède. 1749–1804. *Histoire Naturelle, générale et particulière, avec la description du Cabinet du Roi*. 36 vols. Paris: L'Imprimerie Royale.

Maupertuis, P. L. 1750. *Essai de Cosmologie*. Leiden: Johann Bernoulli.

Paul, D. B., J. Stenhouse, and H. G. Spencer. 2013. The two faces of Robert FitzRoy, Captain of HMS *Beagle* and Governor of New Zealand. *The Quarterly Review of Biology* 88: 219–25.

Sulloway, F. J. 1982. Darwin and his finches: the evolution of a legend. *Journal of the History of Biology* 15: 1–53.

Chapter 3

Bartholomew, M. 1973. Lyell and evolution: an account of Lyell's response to the prospect of an evolutionary ancestry for man. *British Journal for the History of Science* 6: 261–303.

Browne, J. 1995. *Charles Darwin*, Vol. 1: *Voyaging*. Princeton, NJ: Princeton University Press.

Darwin, C. 1837. Notebook B [1837.07]. (Transcription and apparatus by American Museum of Natural History.) Cambridge: Cambridge University Library. www.amnh.org/research/darwin-manuscripts/catalogue-darwin-manu scripts/cambridge-university-library?darbaseurl=https%3A%2F%2Fdarwin.am nh.org%2Fviewer.php%3Feid%3D73127

Darwin, C. 1842. *The Structure and Distribution of Coral Reefs, Being the first part of the geology of the voyage of the Beagle, under the command of Capt. Fitzroy, R.N. during the years 1832 to 1836*. London: Smith, Elder, and Co.

Darwin, C. 1958. *The Autobiography of Charles Darwin (1809–1882); with the Original Omissions Restored* (Edited and with appendix & notes by N. Barlow). London: Collins.

de Tocqueville, A. 1835. *Democracy in America*. London: Saunders & Otley.

Dempster, W. J. 1996. *Evolutionary Concepts in the Nineteenth Century: Natural Selection and Patrick Matthew*. Edinburgh & Durham: Pentland Press Ltd.

Descartes, R. 1853 [1637]. *Discourse on the Method of Rightly Conducting the Reason and Seeking Truth in the Sciences*. 2nd ed. (Translated by J. Veitch). London: Simpkin, Marshall, and Co.

Desmond, A. and J. Moore. 1991. *Darwin: The Life of a Tormented Evolutionist*. New York: Warner Books.

FitzRoy, R. 1839. *Narrative of the Surveying Voyages of His Majesty's Ships Adventure and Beagle Between the Years 1826 and 1836*. London: Henry Colburn.

Galera, A. 2017. The impact of Lamarck's theory of evolution before Darwin's theory. *Journal of the History of Biology* 50: 53–70.

Gould, J. 1830–32. *A Century of Birds from the Himalaya Mountains*. London: [John Gould].

Huzel, J. P. 2006. *The Popularization of Malthus in Early Nineteenth-Century England: Martineau, Cobbett and the Pauper Press*. Aldershot: Ashgate.

Malthus, T. 1798. *An Essay on the Principle of Population*. London: J. Johnson.

Martineau, H. 1833. *Illustrations of Political Economy; No. 6: Weal and Woe in Garveloch*. Boston, MA: Leonard C. Bowles.

Martineau, H. 1833–34. *Poor Laws and Paupers Illustrated*. London: C. Fox.

Martineau, H. 1834. *Illustrations of Taxation*. London: C. Fox.

Martineau, H. 1837. *Society in America*. London & New York: Saunders & Otley.

Ruse, M. 1979. *The Darwinian Revolution: Nature Red in Tooth and Claw*. Chicago, IL: University of Chicago Press.

Russell, R. 2011. *The Business of Nature: John Gould and Australia*. Sydney: National Library Australia.

van Wyhe, J. 2009. Charles Darwin's Cambridge life, 1828–1831. *Journal of Cambridge Studies* 4: 2–13.

Victor, D. 2022. Stolen Darwin notebooks, missing for decades, are returned. *The New York Times* (5 April). www.nytimes.com/2022/04/05/world/europe/charles-darwin-notebooks-cambridge-library.html

Chapter 4

Anon. [Chambers, R.] 1844. *Vestiges of the Natural History of Creation*. London: John Churchill.

Anon. [Wallace, A. R.] 1869. Sir Charles Lyell on Geological Climates and the Origin of Species. *Quarterly Review* 126, 359–94.

"antitype, n." *OED Online*. June 2022. Oxford University Press. www-oed-com /view/Entry/8913

Bateson, G. 1980. *Mind & Nature: A Necessary Unity*. New York: Bantam.

Brand, S. 1976. For God's sake, Margaret! Gregory Bateson & Margaret Mead. *CoEvolutionary Quarterly* 10: 32–44.

Darwin, C. 1859. *On the Origin of Species by Means of Natural Selection, or the Preservation of Favoured Races in the Struggle for Life*. London: Murray.

Darwin, C. 1862. *On the Various Contrivances by which British and Foreign Orchids are Fertilised by Insects*. London: Murray.

Darwin, C. 2004 [1871/1879]. *The Descent of Man; or Selection in Relation to Sex*, 2nd ed. New York: Penguin Classics.

Houghton, G. 1882. *Chronicles of the Photographs of Spiritual Beings*. London: E. W. Allen.

Kottler, M. 1985. Charles Darwin and Alfred Russel Wallace: two decades of debate over natural selection. In *The Darwinian Heritage* (Edited by D. Kohn). 367–432. Princeton, NJ: Princeton University Press.

Kutschera, U. 2003. A comparative analysis of the Darwin–Wallace papers and the development of the concept of natural selection. *Theory in Biosciences* 122: 343–59.

Lyell, C. 1830–1833. *Principles of Geology: Being an Attempt to Explain the Former Changes of the Earth's Surface, by Reference to Causes Now in Operation*. 3 vols. London: John Murray.

Peterson, E. L. 2020. What methods do life scientists use? A brief history with philosophical implications. In *Philosophy of Science for Biologists* (Edited by K. Kampourakis and T. Uller). 168–92. Cambridge: Cambridge University Press.

Shermer, M. 2002. *In Darwin's Shadow: The Life and Science of Alfred Russel Wallace*. New York: Oxford University Press.

Slotten, R. A. 2004. *A Heretic in Darwin's Court: The Life of Alfred Russel Wallace*. New York: Columbia University Press.

Wallace, A. R. 1855. On the law which has regulated the introduction of new species. *Annals and Magazine of Natural History* 16 (2nd series). http://people.wku.edu/charles.smith/wallace/S020.htm

Wallace, A. R. 1869. Geological time and the origin of species. *Quarterly Review* 126: 359–94.

Wallace, A. R. 1869. *The Malay Archipelago*, 2nd ed. London: Macmillan

Wallace, A. R. 1870. *Contributions to the Theory of Natural Selection*. London: Macmillan & Co.

Wallace, A. R. 1875. A defense of modern Spiritualism. *On Miracles and Modern Spiritualism*. London: James Burns.

Wallace, A. R. 1876. *The Geographical Distribution of Animals; With a Study of the Relations of Living and Extinct Faunas as Elucidating the Past Changes of the Earth's Surface*. New York: Harper & Brothers.

Chapter 5

Anon. [Chambers, R.] 1844. *Vestiges of the Natural History of Creation*. London: John Churchill.

Coleman, W. 1971. *Biology in the Nineteenth Century: Problems of Form, Function, and Transformation*. London: John Wiley & Sons, Inc.

Darwin, C. 1859. *On the Origin of Species by Means of Natural Selection, or the Preservation of Favoured Races in the Struggle for Life*. London: Murray.

Darwin, C. 1959 [1859–1872]. *The Origin of Species; a Variorum Text* (Edited by M. Peckham). Philadelphia: University of Pennsylvania Press.

Darwin, C. 1865. The movements and habits of climbing plants. *Journal of the Linnean Society of London (Botany)* 9: 1–118.

Darwin, C. 1885 [1868]. *The Variation of Animals and Plants Under Domestication*, 2nd ed. London: John Murray.

Darwin, C. 2004 [1871/1879]. *The Descent of Man; or Selection in Relation to Sex*, 2nd ed. New York: Penguin Classics.

Darwin, C. 2009 [1872/1890]. *The Expression of the Emotions in Man and Animals*, 2nd ed. New York: Penguin Classics.

Darwin, C. 1875. *Insectivorous Plants*. London: John Murray.

Darwin, C. 1876. *The Effects of Cross and Self Fertilisation in the Vegetable Kingdom*. London: John Murray.

Darwin, C. 1877. *The Different Forms of Flowers on Plants of the Same Species*. London: John Murray.

Darwin, C. 1880. *The Power of Movement in Plants*. London: John Murray.

Darwin, C. 1881. *The Formation of Vegetable Mould Through the Action of Worms*. London: John Murray.

Darwin, E. 1804. *The Temple of Nature; or, The Origin of Society*. American ed. Baltimore, MD: Butler and Bonsal & Niles.

Darwin, E. 1809 [1794–96]. *Zoonomia: or the Laws of Organic Life*, Vol. 1, 3rd American ed. Boston, MA: Thomas & Andrews.

Desmond, A. J. 1989. *The Politics of Evolution: Morphology, Medicine, and Reform in Radical London*. Chicago: University of Chicago Press.

Endersby, J. 2003. Darwin on generation, pangenesis, and sexual selection. In *The Cambridge Companion to Darwin* (Edited by J. Hodge and G. Radick). 69–91. Cambridge: Cambridge University Press.

Ghiselin, M. T. 1969. *The Triumph of the Darwinian Method*. Berkeley, CA: University of California Press.

Haldane, J. B. S. 1959. Natural selection. In *Darwin's Biological Work: Some Aspects Reconsidered* (Edited by P. R. Bell). 101–49. Cambridge: Cambridge University Press.

Newton, I. 1687. *Philosophiae Naturalis Principia Mathematica*. London: S. Pepys.

Ospovat, D. 1981. *The Development of Darwin's Theory: Natural History, Natural Theology, and Natural Selection, 1838–1859*. Cambridge: Cambridge University Press.

Richards, R. J. 1987. *Darwin and the Emergence of Evolutionary Theories of Mind and Behavior*. Chicago, IL: University of Chicago Press.

Ruse, M. 1979. *The Darwinian Revolution: Science Red in Tooth and Claw*. Chicago, IL: University of Chicago Press.

Sober, E. 2011. *Did Darwin Write the "Origin" Backwards? Philosophical Essays on Darwin's Theory*. Buffalo, NY: Prometheus.

Wallace, A. R. 1864. The origin of human races and the antiquity of man deduced from the theory of "Natural Selection" (2010). *Alfred Russel Wallace Classic Writings*. Paper 6. https://digitalcommons.wku.edu/dlps_fac_arw/6

Chapter 6

The quotes from Darwin's autobiography are available at http://darwin-online.org.uk.

Browne, J. 2002. *Charles Darwin*, Vol. 2: *The Power of Place*. Princeton, NJ: Princeton University Press.

Büchner, L. 1855. *Kraft und Stoff [Force and Matter]*. Frankfurt (Main): Meidinger Sohn.

Büchner, L. 1857. *Natur und Geist [Nature and Spirit]*. Frankfurt (Main): Meidinger Sohn.

Büchner, L. 1869. *Die Stellung des Menschen in der Natur [Man's Place in Nature]*. Leipzig: T. Thomas.

Clifford, W. K. 1879. The ethics of belief. In *Lectures and Essays*, Vol. II. London: Macmillan.

Cooke, J. P. 1864. *Religion and Chemistry; or, Proofs of God's Plan in the Atmosphere and its Elements*. New York: Charles Scribner.

Davis, T. 2016. The evolution of Darwin's religious faith. *BioLogos*. https://biologos.org/articles/the-evolution-of-darwins-religious-faith/

Darwin, C. 1859. *On the Origin of Species by Means of Natural Selection, or the Preservation of Favoured Races in the Struggle for Life*. London: Murray.

Darwin, F. 1929. The religion of Charles Darwin. In *Autobiography of Charles Darwin* (Edited by F. Darwin). London: Watts & Co.

Dawkins, R. 1986. *The Blind Watchmaker: Why the Evidence of Evolution Reveals a Universe Without Design*. New York: W. W. Norton.

Desmond, A. and J. Moore. 1991. *Darwin: The Life of a Tormented Evolutionist*. New York: Warner Books.

Fiske, J. 1874. *Outlines of Cosmic Philosophy: Based on the Doctrine of Evolution, with Criticisms on the Positive Philosophy*. 2 vols. London: Macmillan and Co.

Fullerton, W. Y. 1930. *J. W. C. Fegan: A Tribute*. London: Marshall, Morgan, & Scott, Ltd.

Innes, J. B. 1882. [Recollections of Charles Darwin]. CUL-DAR112.B85–B92 (Edited by John van Wyhe) *Darwin Online*. http://darwin-online.org.uk/con tent/frameset?itemID=CUL-DAR112.B85-B92&viewtype=text&pageseq=1

Keynes, R. 2001. *Annie's Box: Charles Darwin, His Daughter, and Human Evolution*. London: Fourth Estate.

Lady Hope. 1915. Darwin & Christianity. *The Watchman-Examiner* [Boston], n.s.3 (19 August): 1071.

LeDrew, S. 2016. *Evolution of Atheism: The Politics of a Modern Movement*. New York: Oxford University Press.

Moore, J. R. 1985. Darwin of Down: the evolutionist as squarson–naturalist. In *The Darwinian Heritage* (Edited by D. Kohn). 435–81. Princeton, NJ: Princeton University Press.

Moore, J. R. 1994. *The Darwin Legend*. Grand Rapids, MI: Baker Books.

James, W. 1956 [1896]. The will to believe. In *The Will to Believe and Other Essays in Popular Philosophy*. New York: Dover Publications.

Marston, P. (Unpublished, 2002). Charles Darwin and Christian faith. www .paulmarston.net/papers/scienceandfaith/Darwin%20and%20Christian%20Fa ith.pdf

Paylor, S. 2005. Edward B. Aveling: the people's Darwin. *Endeavour* 29: 66–71.

Physicus [Romanes, G. J.] 1878. *A Candid Examination of Theism*. London: Trübner & Co.

Chapter 7

Beddoe, J. 1862. *The Races of Britain: A Contribution to the Anthropology of Western Europe*. Bristol/London: Arrowsmith & Trübner.

Browne, J. 2002. *Charles Darwin: The Power of Place*. Princeton, NJ: Princeton University Press.

Biddiss, M. D. 1997. History as destiny: Gobineau, H. S. Chamberlain and Spengler. *Transactions of the Royal Historical Society* 7: 73–100.

Carlson, E. A. 2001. *The Unfit: A History of a Bad Idea*. Cold Spring Harbor, NY: Cold Spring Harbor Press.

Crook, P. 1994. *Darwinism, War and History: The Debate over the Biology of War from the 'Origin of Species' to the First World War*. Cambridge: Cambridge University Press.

Darwin, C. 1845. *Journal of Researches into the Natural History and Geology of the Countries Visited During the Voyage of H.M.S. Beagle*. 2nd ed. London: John Murray.

Darwin, C. 1859. *On the Origin of Species by Means of Natural Selection, or the Preservation of Favoured Races in the Struggle for Life*. London: Murray.

Darwin, C. 1885 [1868]. *The Variation of Animals and Plants Under Domestication*, 2nd ed. London: John Murray.

Darwin, C. 2004 [1871/1879]. *The Descent of Man; or Selection in Relation to Sex*. 2nd ed. New York: Penguin Classics.

Darwin, C. 2009 [1872/1890]. *The Expression of the Emotions in Man and Animals*. 2nd ed. New York: Penguin Classics.

Davis, J. B. 1865. *Crania Britannica: Delineations and Descriptions of the Skulls of the Aboriginal and Early Inhabitants of the British Islands*. London: Taylor & Francis.

Gobineau, J. A., COMTE de. 1853–55. *Essai sur l'inégalité des races humaines [Essay on the Inequality of Human Races]*. Paris: Firmin Didot.

Gould, S. J. 1981. *The Mismeasure of Man*. New York: W. W. Norton.

Grant, M. 1916. *The Passing of the Great Race, or, the Racial Basis of European History*. New York: Charles Scribner's Sons.

Greg, W. R. 1868. On the failure of "natural selection" in the case of man. *Frasier's Magazine* 78: 354–55.

Hartmann, B. 1997. Population control I: birth of an ideology. *International Journal of Health Services: Planning, Administration, Evaluation* 27: 523–40.

Hofstadter, R. 1955. *Social Darwinism in American Thought*. Revised ed. Boston, MA: Beacon.

Huxley, T. H. 1897. Evolution and ethics (1893). *Evolution and Ethics and Other Essays*. New York: D. Appleton & Co.

Kellogg, V. 1907. *Darwinism To-Day*. New York: H. Holt & Co.

Kellogg, V. 1917. *Headquarters Nights: A Record of Conversations and Experiences at the Headquarters of the German Army in France and Belgium*. London: Euston Grove Press.

Largent, M. A. 1999. Bionomics: Vernon Lyman Kellogg and the defense of Darwinism. *Journal of the History of Biology* 32: 465–88.

Morel, B. A. 1857. *Traité des Degenerescences Physiques, Intellectuelles et Morales de l'espece Humaine [Treatise on Human Physical, Intellectual, and Moral Degeneration]*. Paris: Masson.

Nott, J. C., G. R. Gliddon, S. G. Morton, et al. 1854. *Types of Mankind: Or, Ethnological Researches Based Upon the Ancient Monuments, Paintings, Sculptures, and Crania of Races, and Upon Their Natural, Geographical, Philological and Biblical History*. Philadephia: Lippincott, Grambo & Co.

O'Connell, J. and M. Ruse. 2021. *Social Darwinism*. Cambridge: Cambridge University Press.

Olusoga, D. and C. W. Ericksen. 2011. *The Kaiser's Holocaust: Germany's Forgotten Genocide and the Colonial Roots of Nazism*. New York: Faber & Faber.

Peterson, E. [Forthcoming 2023]. Myth 22: That Darwin's hatred of slavery reflected his beliefs in racial equality. In *Darwin Mythology: Debunking Myths, Correcting Falsehoods* (Edited by K. Kampourakis). Cambridge: Cambridge University Press.

Quêtelet, A. 1835. *Sur l'homme et le développement de ses facultés, ou essai de physique sociale [On Man and the Development of His Abilities, an Essay on Social Physics]*. Paris: Bachelier.

Radick, G. 2018. How and why Darwin got emotional about race. In *Historicizing Humans: Deep Time, Evolution, and Race in Nineteenth-Century British Sciences* (Edited by E. Sera-Shriar). Pittsburgh, PA: University of Pittsburgh Press.

Richards, R. J. 2013. *Was Hitler a Darwinian? Disputed Questions in the History of Evolutionary Theory*. Chicago, IL: University of Chicago Press.

Ruse, M. 1979. *Sociobiology: Sense or Nonsense?* Dordrecht, Netherlands: Reidel.

Virey, J.-J. 1801. *Histoire naturelle du genre humain [The Natural History of Humanity]*. Paris: Dufart.

Concluding Remarks

Browne, J. 1982. New developments in Darwin studies? *Journal of the History of Biology* 15: 275–80.

Browne, J. 2022. Reflections on Darwin historiography. *Journal of the History of Biology* 55: 381–93.

Dawkins, R. 1976. *The Selfish Gene*. New York: Oxford University Press.

Desmond, A. and J. Moore. 1991. *Darwin: The Life of a Tormented Evolutionist*. New York: Warner Books.

Flannery, M. C. 2006. The Darwin industry. *The American Biology Teacher* 68: 163–66.

Fuentes, A. 2021. *The Descent of Man*, 150 years on. *Science* 372: 769. www.science.org/doi/10.1126/science.abj4606

Greene, J. C. 1975. Reflections on the progress of Darwin studies. *Journal of the History of Biology* 8: 243–73.

Greene, J. C. 1981. *Science, Ideology, and World View: Essays in the History of Evolutionary Ideas*. Berkeley, CA: University of California Press.

Hofstadter, D. 1979. *Gödel, Escher, Bach: an Eternal Golden Braid*. New York: Basic Books.

Holton, G. 1977. Sociobiology: the new synthesis? *Newsletter on Science, Technology, & Human Values* 21: 28–43.

Lenoir, T. 1987. The Darwin industry. *Journal of the History of Biology* 20: 115–30.

Mendelsohn, E. 1967. Review of: *History of Science. An Annual Review of Literature, Research and Teaching*. Vol. 4, 1965 by A. C. Crombie, M. A. Hoskin. *Bulletin of the History of Medicine* 41: 183–84.

Midgely, M. 1985. *Evolution as a Religion: Strange Hopes and Stranger Fears*. London: Methuen & Co.

Moore, J. 1984. On revolutionizing the Darwin industry: a centennial retrospect. *Radical Philosophy* 9: 13–22.

Ospovat, D. 1981. *The Development of Darwin's Theory: Natural History, Natural Theology, and Natural Selection, 1838–1859*. Cambridge: Cambridge University Press.

Ruse, M. 1974. The Darwin industry – a critical evaluation. *History of Science* 12: 43–58.

Ruse, M. 1996. *Monad to Man: The Concept of Progress in Evolutionary Biology*. Cambridge, MA: Harvard University Press.

Tagliabue, J. 1996. Pope bolsters Church's support for scientific view of evolution. *The New York Times* (25 October), 1A. www.nytimes.com/1996/10/25/world/pope-bolsters-church-s-support-for-scientific-view-of-evolution.html

Time staff. 1977. Why you do what you do – Sociobiology: a new theory of behavior. *Time Magazine* (1 August). https://content.time.com/time/subscriber/article/0,33009,915181,00.html

Whiten, A., W. Bodmer, B. Charlesworth, et al. 2021. RE: '*The Descent of Man*,' 150 years on. Jun 6 comment to Fuentes 2021. www.science.org/doi/10.1126/science.abj4606

Wilson, E. O. 1975. *Sociobiology: The New Synthesis*. Cambridge, MA: Harvard University Press.

Figure Credits

Figure 1.1   Stipple engraving by J. Heath, 1804, after J. Rawlinson. Wellcome Collection. Public domain.

Figure 1.2   Wikimedia commons. Public domain.

Figure 1.3   Wellcome Collection. Public domain.

Figure 1.4   Wellcome Collection. Public domain.

Figure 1.5   Wellcome Collection. Public domain.

Figure 1.6   Wikimedia commons. Public domain.

Figure 2.1   www.asap.unimelb.edu.au/bsparcs/covingto/gifs/covingto.jpg

Figure 2.2   Reproduced from Lamarck, *Philosophie Zoologique* (1809).

Figure 2.3   Wellcome Collection. Public domain.

Figure 3.1   Photo by Mary Evans, Norwegian Digital Learning Arena (https://ndla.no/article/24544). CC BY-NC 4.0.czQ

Figure 3.2   Wood engraving. Wellcome Collection. Public domain.

Figure 3.3   Author illustration.

Figure 3.4   Wikimedia commons. Public domain.

Figure 3.5   Reproduced from Darwin's "Notebook B" (July 1837), by kind permission of the Syndics of Cambridge University Library.

Figure 4.1   Wellcome Collection. Public domain.

Figure 4.2   Wikimedia commons. Public domain.

Figure 4.3   Wellcome Collection. Public domain.

Figure 4.4   Wikimedia commons. Public domain.

Figure 4.5   Wikimedia commons. Public domain.

Figure 4.6   Author photograph in 2017.

Figure 4.7   Author photograph in 2015.

Figure 4.8   Reproduced from Houghton, *Chronicles of the Photographs of Spiritual Beings* (1882). Public domain.

Figure 5.1   Reproduced from Darwin, *On the Origin of Species* (1859), insert between pages 116 and 117.

Figure 5.2   Reproduced from Darwin, *Variation in Animals and Plants Under Domestication* (1868).

Figure 5.3   Royal Institute for NEET/ IIT-JAM. https://royalinstitute10163 .blogspot.com/

Figure 5.4   Reproduced from Darwin, *Descent of Man; and Selection in Relation to Sex* (1871).

Figure 6.1   Wikimedia commons. Public domain.

Figure 6.2   Author photograph in 2015.

Figure 6.3   National Library of Wales. Public domain.

Figure 6.4   Wikimedia Commons. Public domain.

Figure 7.1   mikroman6 / Moment / Getty Images.

Figure 7.2   Frontispiece Vol. 1, *Treasury of Human Inheritance*, edited by Karl Pearson, London, Dulau and Co, 1912. License: In copyright. https://wellcomecollection.org/works/t3s35vtf

Figure 7.3   Bundesarchiv, Bild 102–16748 / Georg Pahl / CC-BY-SA 3.0, CC BY-SA 3.0 DE, via Wikimedia Commons.

Figure C.1   ullstein bild / Contributor / Getty Images.

Index

Page numbers for figures are in *italics*

adaptation to environment, 121
  Galápagos Archipelago, 34, 47, 66, 107
  Maupertuis, 35
  Cuvier, 42, 45–46
agnosticism, Darwin's, 141, 143–44, 180
The *Allmacht* (supreme power or principle) of natural selection, 87, 88, 148–51
analogy, Darwin's use, 99–101, 103, 106–7, 179
anatomy, comparative, 42
animal breeding *see* domestic breeding
Annals and Magazine of Natural History, 70
antisemitism, 157
anti-slavery movement, 6–7, 29–30, 132, 152–53, 158–60
antitype, 76
argument from design, 20, 36, 132–33, 141
Aristotle, 34, 99–100
Aryan race science, 159–61, 166–67
atheism, Darwin's, 4, 127–28, 139–41, 150–51, 175, 180
Aveling, Edward B., 139–41

Babbage, Charles, 57–58
barnacles, 3, 58–59, 80, 85, 101–3, 120
Bateson, Gregory, 95–96
Bateson, William, 95
The Beagle voyage, 21–30, 46–50, 59
beaks in birds, variation, 7, 15–16, 46, 48, 52–53, 107
Beddoe, John, 156–57
beetles, 19, 77, 178
behavioral homologies, 122–24
belief by proxy, 132–39
belief in God *see* religious beliefs
  Big Idea, Darwin's, 4, 66, 98, 121, 179–80
big science, 125
biology, first use of term, 36–37
Blends of Pathological Anthropology and Mental Medicine (Morel), 160–61
Bonnet, Charles, 35
The Botanic Garden (E. Darwin), 7
Bowler, Peter, 173–74
Brack, Dr. Viktor H., 167
brain, human, 89–90, 123–24, 156–57
Brooke, Sir James, 67–68
Browne, Janet (biographer of Darwin), 175–76

Bryan, William Jennings, 147, 150–51
Büchner, Ludwig, 139–41, 149–50
Buckland, William, 44–45
Buffon, Comte de, 35–36
Burton, John Barlow, 76
Bynow, Benjamin, 26

Cambridge University, 12, 53–54
    Darwin archive, 170, 173
A Candid Examination of Theism
    (Romanes), 141
Carlyle, Thomas, 57–58
catastrophism, 44–45
A Century of Birds from the Himalaya
    Mountains (Gould), 52
Christ's College, Cambridge, 12, 53
church supporting, 132–39, 180
class distinctions, Darwin and Wallace,
    87–94
Clifford, W. K., 134
co-evolution, 108
collector vs. naturalist, distinction
    between Wallace and Darwin, 76, 85
Colnett, Captain James, 49
colonialism, 157
common ancestry, organisms, 153
common descent with modification *see*
    descent
comparative anatomy, 42
complex behaviors, evolution, 9
concentration camps, 148; *see also* death
    camps
Cooke, Josiah P., 132–33
coordination of parts argument, 107, 115
corals/coral reefs, 29, 46, 58–59
cosmism, 142
Cosmos (von Humboldt), 27
Covington, Syms, 28–29, *34*
Coxon, Elizabeth, 52
Crania Britannica (Davis), 156–57
creationism, 59–60, 76

Fitzroy, 30
Cuvier, 44
    intelligent design, 175
    "scientific," 145, 172–73, 174
Crichton-Browne, J., 123–24
culture wars, 172–73
Cuvier, Georges, 42, *43*, 46
    coordination of parts argument,
        107, 115

Darrow, Clarence, 147
Darwin, Annie (daughter), 133, 143, *144*
Darwin, Charles, *127*; *see also* Darwinism
    Big Idea, 4, 66, 98, 121, 179–80
    colonialism, racism and
        xenophobia, 168
    family history, 9–10
    first-cousin marriage, 59, 80
    his Darwinism, 124–25
    late books, 124–25
    legacy and legends, 170–78
    misconceptions concerning ideas, 1–4,
        32–34, 47, 49–50, 86–87, 127–28,
        179–81
    morality, 168–69
    slowness to publish, 78
    support for Wallace, 78
Darwin, Charles Waring (son), 80
Darwin, Erasmus (brother), 12, 54, 56
Darwin, Erasmus (grandfather), 6, 36,
    62–63
    common descent with modification,
        41–42, 50
    evolutionary ideas, 5–9, 62–63,
        124–25, 179
    family, 9–10
    sexual selection, 120–22
Darwin, Francis (son), 1–2, 131
Darwin, Major Leonard (son), 88
Darwin, Robert Waring (father), 9–10
Darwin Day, 2–3

Darwin Fox, William (cousin), 19
"Darwin industry," 3, 4, 170–71, 175–76
Darwin: The Life of a Tormented Evolutionist (Desmond and Moore), 174
The Darwinian Heritage (Kohn), 173
Darwinism, supposed consequences in twentieth century, 147–48
    death camps, 158–69
    German Social Darwinism, 148–51
    survival of the fittest applied to humans, 152–57
Darwinism To-Day (Kellogg), 149
Davenport, Charles, 166
Davis, Joseph B., 156–57
Dawkins, Richard, 172–73
de Maillet, Benoit, 35
de Tocqueville, Alexis, 54–56
death camps, 4, 158–69, 180–81
"deathbed conversion," Darwin's, 126–27, 128–31
deduction vs. induction as scientific method, 112–13
Democracy in America (de Tocqueville), 54–56
Descartes, Réne, 34–35
descent/common descent with modification, 7, 32–34
    The Beagle journal, 47, 48, 50
    Darwinism, 124–25
    early evolutionists, 36, 41–42
    Grant, 16
    of man, 8, 89, 118–22
    modified minds as well as bodies, 9, 122–24
    "Notebook B," 62–63
    pigeon breeding, 115
    through natural selection, 85, 98, 111, 113–14
    variation under nature, 101–2
    Wallace, 90–91

The Descent of Man (C. Darwin), 9, 86, 89, 97, 118–22, 156–57, 177–78
design by God *see* argument from design
Desmond, Adrian (biographer), 174
dimorphism, sexual, 119–20, *120*
Discourse on the Method (Descartes), 35
divergence, 104, 115, 117–18
divergence diagram, 108–11, *109*
diversity vs. resource competition, 105
domestic breeding, 41, 86, 114–18
    Darwin's use as analogy, 99–101
    pigeons, 115, *116*
    poultry, 76–77
dominance leading to dominance, 156
Down Friendly Society, 134–36
Down House, 1, 83–84, *83*, *84*, 128, 129, *130*
Downe, St. Mary's church, 134–36
duty, doing one's, 133–34, 143–44, 151, 168–69, 178
Dyak people of Borneo, 69–70

Earle, Augustus, 26
earthworm stone, 1–2, 176, 177
economic policies, 88
Edinburgh transformists, 39
Edinburgh University, 12
Edmonstone, John, 13–16
educational distinctions, Darwin and Wallace, 87–94
embryological development, 41
emotional expression, 122–24, 153, 179–80
Essay on Cosmology (Maupertuis), 35
Essay on Population (Malthus), 54
Essay on the Inequality of the Human Races (Gobineau), 159–60
Essays in Contributions to the Theory of Natural Selection (Wallace), 89–90
eugenics, 9–10, 88, 163–65, 167, 173–74
    USA legislation, 166

Eugenics Education Society, London, 164
Eugenics Records Office, 166
Evolution and Ethics (Huxley), 169
evolutionary tree, 63–65
   C. Darwin, *64*, 66
   Lamarck, *38*, 63–65
   Wallace, 75
evolution/evolutionary ideas
   1794–1835, 5–11
   The Beagle voyage, 21–30
   Darwin at University, 18–21
   first use as biological concept,
     35
   human, 8, 89, 118–22
   the leisure class, 11–18
   pre-dating C. Darwin, 5–9, 34–41,
     62–63, 124–25, 179
   related to survival or not, 87
   religious beliefs, 174
   tempo, 86
*The Expression of the Emotions in Man
   and Animals* (C. Darwin), 9, 97,
   122–24, 152–53, 179–80
extinctions, mass, 44–45

family history, 9–10
Fegan, Rev. J. W. C., 137–38
Fellow of the Royal Society, 59
"Final Solution," 167
finches, Galápagos, 48, 52, *53*,
   58–59, 107
first-cousin marriages, 59, 80
Fiske, John, 142
Fitzroy, Capt. Robert, 21–26, *25*,
   44–45
   Galápagos, 47
   tortoise migration, 49–50
fossils, 7, 26–27, 41, 44
Frankenstein (Shelley), 37
Fuegians, 22–24, *23*, 27–28
Fuentes, Agustin, 177–78

Galápagos Archipelago, 3, 4, 30, 31–32
   Alfred Russel Wallace, 74–75
   Darwin's visit, 32–34, 46–50, 179
   finches, 48, 52, *53*, 59, 107
Galton, Sir Francis (cousin), 9–10, 88,
   163–65
gemmules, 116–18, 124–25, 179–80
Genesis, book of, 128–29
genocide, 154–56, 158–69
gentleman naturalists, 19, 21–22, 59–60,
   82, 85, 125, 173
Geographical Distribution of Animals
   (Wallace), 94–95
Geological Society of London, 59, 83–84
Geological Time and the Origin of Species
   (Wallace), 89
geological view, 21–22, 26–27, 40–41,
   113, 144–46
   C. Darwin, 46–47, 59
German Social Darwinism, 148–51
germ-plasm theory, *117*
Gobineau, Arthur de, 159–60
Goethe, Johann Wolfgang van, 45–46
Gould, John, 46, 51–52, 107
gradualism, 86
Grant, Madison, 166
Grant, Robert E., 16, *17*, 62–63
Gray, Asa, 80, *82*, 143
great man theory, 170–71
Greg, William R., 163
gulf stream, 49–50

Haeckel, Ernst, 95, 149–50
Headquarters Nights (Kellogg), 149
Hellyer, Edward H., 26
Henslow, Rev. John S., 20–22, 44–45, 53
Hobbes, Thomas, 151
Hofstadter, Richard, 162
Hooker, Sir Joseph D., 80, *81*
Hope, Admiral James, 130–31

Hope, Lady, 126, 128–31
Horsman, Rev. S. J. O., 136–37
Hudson, Frederick A., 91–92, 93
human brain, 89–90, 123–24, 156–57
human evolution, 8, 89, 118–22
human exceptionalism, 71–72, 88–90
Human Genome Project, 175
human races/breeds/species, 119,
      158–60, *159*
humans as primates, 36, 89–90,
      158
Humboldt, Alexander von, 27
Hutton, James, 26–27
Huxley, Thomas H., 50, 72, 80, *81*,
      117–18, 169
hypnosis, 91

iconic status, 176–78
idiosyncratic approach, 176
iguanas, 33–34
Illustrations of British Entomology
      (Stephens), 19
Illustrations of Political Economy
      (Martineau), 54–57
Illustrations of Taxation (Martineau),
      54–56
induction vs. deduction as scientific
      method, 112–13
inheritance of acquired characteristics,
      37–38, 112–14, 116–18, 179–80
inheritance of variations, 115, 117–18
Innes, Rev. J. B., 135–36
intellectual gap, apes/humans, 123
intelligent design, 175; *see also* argument
      from design
interbreeding, human races, 159–60
inter-/intra-species competition, 87
Introduction of New Species (Wallace),
      70, 71
invertebrates, first use of term, 36–37
isolation, effects on evolution, 113

James, William, 134
Jameson, Robert, 16, 39
Jenyns, Leonard, 57–58
Johnson–Reed Act, 166–67
Journal of Natural History, 70
Journal of Researches (C. Darwin), 59, 74,
      85, 152–53
jungle fowl, 76

Kames, Lord, 5, 158
Kant, Immanuel, 158–60
Kellogg, Dr. Vernon, 148–49, 157
Kew Botanical Gardens, 80, 86–87
Kingsley, Rev. Charles, 155
Kitzmiller v. Dover Area School District
      legal case, 175

Lamarck, Jean-Baptiste, 16, 36–38
   evolutionary tree, *38*
   transmutationism, 115–16
Lamarckism, 39, 80–82, 171–72
Laughlin, Harry H., 166–67
Leclerc, Georges-Louis, 35–36
Lincecum, Dr. Gideon, 163, 165
Lincoln, Abraham, 11–12
Linnean Society of London, 85, 95
London years, Darwin's (1836–42),
      51–53
   career progression, 53–58
   original thinking, 62–66
   researches, 58–62
Lunar Society of Birmingham, 6–7
Lyell, Charles, *60*
   correspondence with Darwin and
      Wallace, 78–80
   human exceptionalism, 71–72
   influence on Darwin, 53–54, 60–62,
      153–55
   Principles of Geology, 26–27, 44–45
   racism, 153–55
   support for Wallace, 78

transmutationism, 39
uniformitarianism, 113

Malay Archipelago (Wallace), 89
Malthus, Rev. Thomas, 54, 56–57, 58, 70–71
    Malthusian theory, 56–58
    population checks, 115
    population principle, 104
    struggle for existence, 103–5
"man is wolf to man," 151
Marriage, Darwin's, 59
Martineau, Harriet, 54–58, *55*, 103–5
mass extinctions, 44–45
materialism, 136, 150–51
Maupertuis, Pierre Louis, 35, 117
Mayr, Ernst, *171*, 171–72
McCormick, Robert, 26
McLean v. Arkansas legal case, 172–73
Mead, Margaret, 96
Meckel–Serres Law, 39
Mendelsohn, Everett, 170
mental illness research, 123–24, 161
misconceptions/misunderstandings, 1–4, 178
    atheism, 127–28, 180
    Darwin vs. Fitzroy, 48
    Darwinism vs. Wallaceism, 86–87, 179
    Darwin's Big Idea, 98, 108, 179–80
    Galápagos, 32–34, 47, 179
    genocide/death camps, 148, 180–81
    natural theology, 20
missionaries, 138–39
mockingbirds, 75
Model Sterilization Law (USA), 166–67
modern synthesis, 171–72
modernism, 136
Monboddo, Lord, 5, 158
monkeys, 8
monogenism vs. polygenism, 5, 119, 158, 180–81

Moore, James (biographer), 129, 174
moral education, 164
moral illnesses, 161
moral principles, 134
Morel, Dr. Bénédict-Augustin, 160–62
moths of Borneo, 68–69
Mount Serembu, 67–68
"The Mount," Shrewsbury (Darwin's boyhood home), *11*
The Movements and Habits of Climbing Plants (C. Darwin), 97

Nash, Wallis and Louisa, 137
natural history, 2, 16–18, 40–41; *see also* gentleman naturalists
Natural History (Buffon), 35–36
The Natural History of Humanity (Virey), 158
natural selection, 4, 97–98
    Darwin on, 85, 87, 105–8, 173–74, 179–80
    Darwin's Darwinism, 124–25
    Darwin's divergence diagram, 108–11
    The Descent of Man, 118–22
    domestic breeding, 114–18
    The Expression of the Emotions in Man and Animals, 122–24
    On the Origin of Species, 98–108
    problems, 112–14
    Wallace, 84–85, 87, 88, 179–80
natural theology, 20–21, 121, 132, 141, 173–74
Natural Theology (Paley), 20
naturalist vs. collector, distinction between Wallace and Darwin, 76, 85
Naziism, 167
neo-Darwinian synthesis, 171–72
neo-Darwinism, 149, 173–74
neutral changes in survival, 107
Nevill-Rolfe, Sybil K., 164
Newton, Isaac, 100–1, 103, 110

"Notebook B" (C. Darwin), 62–63, *64*, 85–86, 105–8

Nott, Josiah C., 159

On Miracles and Modern Spiritualism (Wallace), 94

On Population (Malthus), 58

On the Origin of Species by Means of Natural Selection (C. Darwin), 21, 41, 85–86, 97–98, 113–14, 153
   Chapter 1: Variation Under Domestication, 99–101
   Chapter 2: Variation Under Nature, 101–3
   Chapter 3: Struggle for Existence, 103–5
   Chapter 4: Natural Selection, 105–8
   racial competition, 156

On the Tendency of Varieties to Depart Indefinitely from the Original Type (Wallace), 78–80

On the Various Contrivances by which British and Foreign Orchids are Fertilised by Insects (C. Darwin), 97

orangutans of Borneo, 69

Ospovat, Dov, 173–74

Outlines of Cosmic Philosophy (Fiske), 142

Owen, Richard, 45–46, 53–54, 59–60

Paley, William, 20, 132

pan-Aryanism, 166–67

pangenesis, 35, 116–18, 141, 179–80

Parslow, Joseph (butler), 131

pigeon breeding, 115, *116*

plant breeding *see* domestic breeding

plants, Darwin's application of natural selection, 86–87

Plinian Society, 16–18

polygenism vs. monogenism, 5, 119, 154–55, 158

population feedback loops, 96

poultry, domestication, 76–77

primate evolution, 8, 36

Principes de Philosophie Zoologique (Geoffroy)

Principia Mathematica (Newton), 110

Principles of Geology (Lyell), 26–27, 44–45, 60

Principles of Moral and Political Philosophy (Paley), 20

problems, with Darwin's theory, 112–14

procrastination, 84–85

psychic phenomena, 93

published works on Darwin, 170

Quêtelet, Adolphe, 161

race vs. species, 114

races, human, 119, 158–60, *159*

Races of Britain (Beddoe), 156–57

racism, 151, 153–55

Religion and Chemistry; or, Proofs of God's Plan in the Atmosphere and its Elements (Cooke), 132–33

religious attitudes to evolution, 174

religious beliefs, Darwin's, 4, 126–28, 145–46
   agnosticism, 142–44, 180
   atheism, 139–41
   church supporting, 132–39, 180
   deathbed "conversion," 126–27, 128–31
   geological view, 144–46

resource competition vs. diversity, 105

Romanes, George J., 141

Ruse, Michael, 170–71

Saint-Hilaire, Etienne Geoffroy, 45–46

scientific method, 110, 112–13

Scopes "Monkey Trial," 147

Sedgwick, Rev. Adam, 20–21, 44–45, 53

selection, sexual, 9, 120–22
Selection in Relation to Sex (C. Darwin), 9, 156–57
selection theory, differences between Wallace and Darwin, 86–87
Selfish Gene (Dawkins), 172–73
Serembu, Mount, 67–68
Serres, Antoine Étienne, 39
sexual dimorphism, 119–20, *120*
sexual reproduction, 9
sexual selection, 9, 120–22
Sharp, Harry, 165–66
Shelley, Mary, 37
Shrewsbury School, 12
slavery, 6–7, 29–30, 132, 152–53, 158–60
Social Darwinism, 3, 148, 162
socialism, 136
Society in America (Martineau), 54–56
Sociobiology: The New Synthesis (Wilson), 172–73
Soho Circle, 54, 57–58
species, 40–41; *see also On the Origin of Species*
  definition, 100, 110
  Galápagos finches, 48
  genera, 65–66, 103
  incipient, 102–3
  orangutans of Borneo, 69
  vs. race, 114
  stability/instability, 34–35, 61
  sub-species, 102, 105
  vs. variety, 101–2
  Wallace on, 71–78
speculation vs. evidence, 117–18
Spencer, Herbert, 153
spiritualism, 88–93, 94–96
sterilization, 165–66, 167
Stirling, Bishop, 138–39
The Structure and Distribution of Coral Reefs (C. Darwin), 58–59

struggle for existence, 103–5
The Student's Darwin (Aveling), 140
sub-species, 102, 105
suicides, 24–25
Sulivan, Admiral Sir B. J., 138–39
survival, 87
survival of the fittest, 150, 156, 164, 180–81

taxidermy, 13–16, 51–52
taxonomy, 40–41, 100
temperance movement, 130–31, 134, 137, *138*
The Temple of Nature; or, The Origin of Society (E. Darwin), 8
theology, natural, 20–21, 121, 132, 141, 173–74
Theophrastus, 34
Tierra del Fuego, 27–28
tortoises, 33–34, 49–50
traits *see* variation of traits
Transmutation Notebooks (C. Darwin), 62
transmutation/transmutationism, 5, 41–42, 45–46, 71–72, 80
  French/Scottish, 60–62
  Jean-Baptiste Lamarck, 16, 37–39, 44, 115–16
  Robert Grant, 16–18
tree of life *see* evolutionary tree
Types of Mankind (Nott, Gliddon *et al.*), 151, 154–55, 159

uniformitarianism, 113
university, 12

vaccination, compulsory, 88
The Variation of Animals and Plants Under Domestication (C. Darwin), 97, 99–100, 114–18, 156, 171–72
variation of traits
  behavioral, 122

variation of traits (cont.)
    domestic breeding, 99–101
    natural selection, 106, 114
    sexual selection, 120–22
    species formation, 101–3
Vestiges of the Natural History of Creation
    (Chambers), 71–72, 85, 114–15
A View of the Evidences of Christianity
    (Paley), 20
Virey, Julian-Joseph, 158
Voyage of the Beagle (C. Darwin), 46–47,
    49–50, 74
vulcanism, 16

Wallace, Alfred Russel, 4, 47, 67–71, *68*,
    76–77
    class and educational differences with
        Darwin, 87–94
    vs. Darwin, 71–78, 83–87, 112,
        179
    evolutionary hypothesis, 78–82
    human evolution, 119
    human exceptionalism, 89–90, 149–50
    letter to Darwin, 1858, 70–71
    Spiritualism, 94–96

Wanderings in South America, the North-
    west of the United States, and the
    Antilles (Waterton), 13–16
Warfield, B. B., 143
watchmaker argument, 20
Waterton, Charles, 13–16, *14*
Wedgwood, Emma (wife), 59, 83–84,
    140–41
Wedgwood, Josiah (grandfather),
    6–7, 9–10
Wedgwood, Susannah (mother), 9–10
Wernerian Society, 16–18
Whewell, William, 45–46, 103
white supremacism, 154–56, 158–60
Wilberforce, William, 6–7
Wilson, E. O., 172–73
Wollaston, Thomas V., 80

Zoological Philosophy (Lamarck), 36–37,
    38, 39, 80
Zoological Society of London, 51–52
Zoonomia (Notebook B) (C. Darwin),
    62–63, *64*, 85–86, 105–8
Zoonomia; or the Laws of Organic Life
    (E. Darwin), 7, 16–18, 62–63, 115–16